
Modern Metallography

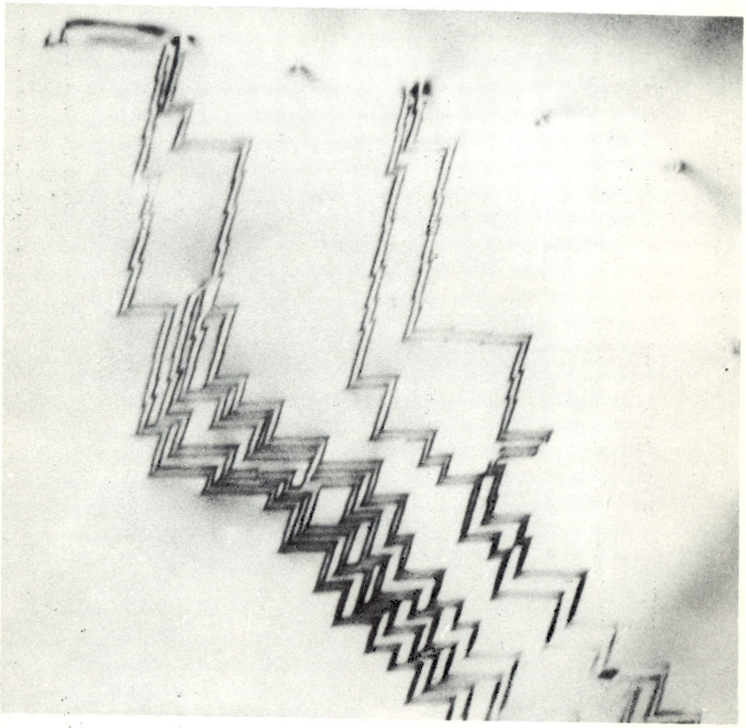

Transmission electron micrograph of antiphase boundaries in rutile.

Modern Metallography

by

R. E. SMALLMAN, B.Sc., Ph.D.
Professor of Physical Metallurgy in the University of Birmingham

and

K. H. G. ASHBEE, B.Sc., Ph.D.
Lecturer in Physics, University of Bristol

THE QUEEN'S AWARD TO INDUSTRY 1966

PERGAMON PRESS
OXFORD · LONDON · EDINBURGH · NEW YORK
TORONTO · SYDNEY · PARIS · BRAUNSCHWEIG

Pergamon Press Ltd., Headington Hill Hall, Oxford
4 & 5 Fitzroy Square, London W.1
Pergamon Press (Scotland) Ltd., 2 & 3 Teviot Place, Edinburgh 1
Pergamon Press Inc., Maxwell House, Fairview Park, Elmsford, New York 10523
Pergamon of Canada Ltd., 207 Queen's Quay West, Toronto 1
Pergamon Press (Aust.) Pty. Ltd., 19a Boundary Street,
Rushcutters Bay, N.S.W. 2011, Australia
Pergamon Press S.A.R.L., 24 rue des Écoles, Paris 5e
Vieweg & Sohn GmbH, Burgplatz 1, Braunschweig

First edition 1966
Reprinted 1969
Library of Congress Catalog Card No. 65-27384

Printed in Great Britain by Hazell Watson & Viney Ltd., Aylesbury, Bucks

08 011570 5 (flexicover)
08 011571 3 (hard cover)

CONTENTS

PREFACE

IT IS now well established that the properties of metals are determined by the presence and behaviour of the defects they contain. These defects include precipitates, cracks, grain boundaries, dislocations, stacking faults and impurity atoms, the theoretical aspects of which have been treated in several books in recent years. There is however, a gap in metallurgical literature for a book suitable for students which deals with the practical techniques of studying defects in metals. It is hoped that the present book will at least partly fill this gap.

The book is primarily written for students of Metallurgy and Materials Science at Universities and Colleges of Advanced Technology, although it is hoped that practising metallurgists will find it useful. Since a comprehensive treatise on metallography would require the writing of several books, we decided to plan the present volume on three main principles: (1) to describe metallographic methods rather than observations; (2) to devote more space to recent methods than to well established ones. It is for this reason that our treatment of, for example, conventional X-ray techniques is by no means detailed; and (3) to consider in more detail the theories which lead to the experimental determination of quantities hitherto unknown, such as stacking-fault energy, Burgers vector, etc.

Finally, we would like to thank many colleagues who have helped during the preparation of this book, particularly those who have supplied original prints of micrographs. Due acknowledgement is given at the appropriate places in the text.

"If they wos a pair o' patent double million magnifyin' gas microscopes of hextra power, p'raps I might be able to see through a flight o' stairs and a deal door; but bein' only eyes, you see, my wision's limited"

(SAM WELLER–*Pickwick Papers*)

CHAPTER I

REFLECTED LIGHT MICROSCOPY

INTRODUCTION

Metallographic examination is usually carried out with the aid of reflected light in order to investigate the nature and distribution of phases in metallurgical specimens, as well as to examine such lattice features as grain boundaries, slip bands, twins and cracks.

SPECIMEN PREPARATION

The specimen for microscopical examination is cut from the bulk metal sample and mounted either in cold-setting plastic or in thermo-setting bakelite using a metallographer's conventional mounting press. The surface to be examined is then ground by hand on a metal file until a suitable flat is obtained, and then on emery papers of decreasing coarseness from 320 grade down to 640 grade. To prevent over-heating, this operation is usually carried out under water or paraffin with the specimen held in such a way that each successive grinding operation is made at right angles to the one before it, and the specimen is not transferred to a new emery paper until the scratches produced by the previous paper have been removed by those of the one in use. A further necessary refinement of the grinding procedure is to wash the specimen carefully with water between successive papers to prevent coarse grit from being carried on to less coarse papers. After grinding on the finest grade emery paper, the specimen should be washed in water and then in alcohol, dried with a hair dryer, and finally polished on a selvyt cloth impregnated with an abrasive (alumina or diamond powder) held in a liquid suspension (water or paraffin). Polishing abrasives of various particle sizes from 10 microns down to $\frac{1}{4}$ micron are commercially available, and the best results are obtained using a series of cloths each impregnated with abrasive of successively finer grade. In almost all cases, a circular motion of the specimen on the selvyt cloth is recommended and is usually accomplished by carrying the selvyt cloth on a mechanical polishing wheel similar to that used by jewellers for lapidary work.

The mechanical polishing treatment involves the removal of surface projections of decreasing coarseness as the sequence file $\longrightarrow$ emery papers $\longrightarrow$ polishing wheel proceeds, which takes place by means of friction between the projections and the grinding or abrading medium. By its very nature, this mechanism produces both heat and plastic deformation in the surface layers of the specimen. In some cases this heat and/or deformation are undesirable, since it alters the thermal and mechanical history of the specimen under examination, but for the majority of specimens the affected surface layers are removed by etching, and hence the surface finally examined is representative of the bulk material.

Chemical and electrolytic polishing methods have been developed for certain metals and, provided a satisfactory surface can be produced, are usually to be preferred to mechanical polishing since the problem of surface damage is avoided. These methods of polishing are of special importance in electron microscopy and will be dealt with later.

THE REFLECTED LIGHT MICROSCOPE

Figure 1 is a ray diagram showing the formation of an image in the compound microscope. The basic components are two convex lenses, called the objective and eyepiece respectively. The object under examination O, is placed in front of the objective at a distance greater than its focal length f_o, but less than $2f_o$. A real inverted and enlarged image I_1, is formed at a distance greater than $2f_o$ from the objective on the side remote from the object. The distance between the objective and the eyepiece is made such that image I_1 is at a distance of less than f_e from the eyepiece, where f_e is the focal length of the eyepiece.

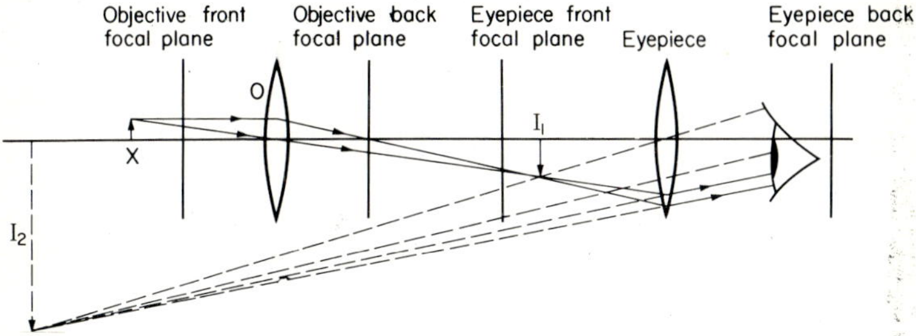

FIG. 1. Ray diagram showing how the image is formed in a compound microscope.

The eyepiece forms a virtual, upright and enlarged image I_2 of I_1, and it is this final image I_2 which is seen by the eye.

With low-power objectives the working distance (i.e. the space between the front surface of the objective and the specimen when the latter is in focus) is large enough to allow light, from a lamp held nearby, to illuminate the specimen. This is known as oblique illumination and is useful in the examination of surface projections (see Chapter III). The working distance is small for high-power objectives and it becomes necessary to illuminate the specimen with light passing vertically down through the objective. The principle of vertical illumination is illustrated in Fig. 2 where light from a source is reflected down through the objective by a half silvered plane mirror held at an angle of 45° to the axis of the microscope. In some microscopes the half-silvered mirror is replaced by a right-angled glass prism which is offset from the axis of the microscope. Light reflected by the specimen and gathered by the objective passes through the half-silvered plane mirror to form the image which is viewed with the eyepiece.

A microscope cannot function properly unless the specimen is correctly illuminated; referring to Fig. 2, so-called "critical illumination" is achieved in the following way.

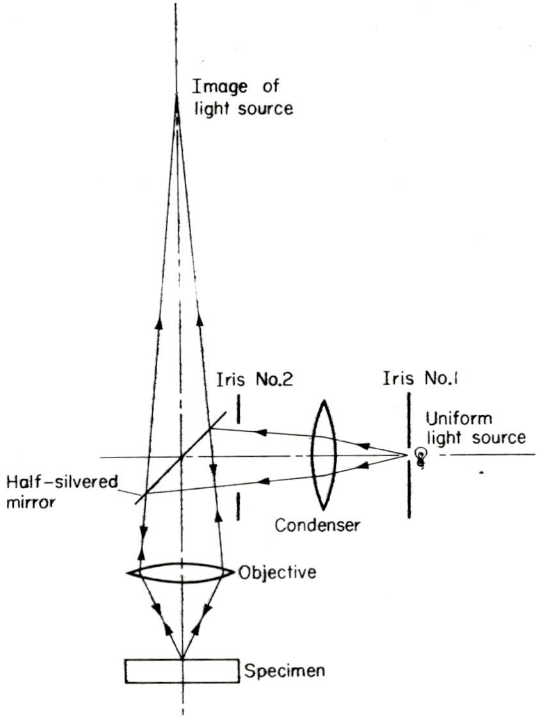

FIG. 2. Vertical illumination of a metallographic specimen.

1. The image of a uniform light source is formed at the specimen surface. The student should check that this is so by removing the microscope eyepiece and adjusting the position of the condenser lens until a sharp image of the lamp filament can be seen.
2. The aperture of No. 1 iris diaphragm is adjusted until it is just too big for the image of its outline to be seen when looking down the microscope. Under these conditions, the illuminated area on the specimen is controlled so that areas of no interest, e.g., the mounting medium, are not

illuminated and cannot scatter light into the objective, thereby "glaring" the image.

3. The aperture of No. 2 iris diaphragm is now decreased until only the area of interest can be seen. This effects an aperture control on the objective and ensures that the least possible amount of the less accurate part of the lens is used.*

For photographic work a more intense light source is

FIG. 3(a). Metallographic photo-microscope. (*Courtesy of Reichert.*)

*N.B. The quality of a lens is inferior at its outer edge than near its centre.

required. In Fig. 3(a) is shown a commercially available photomicroscope, and in Fig. 3(b) a ray diagram, to illustrate how the specimen under examination is illuminated and how the rays reflected by the specimen are received on the photographic plate.

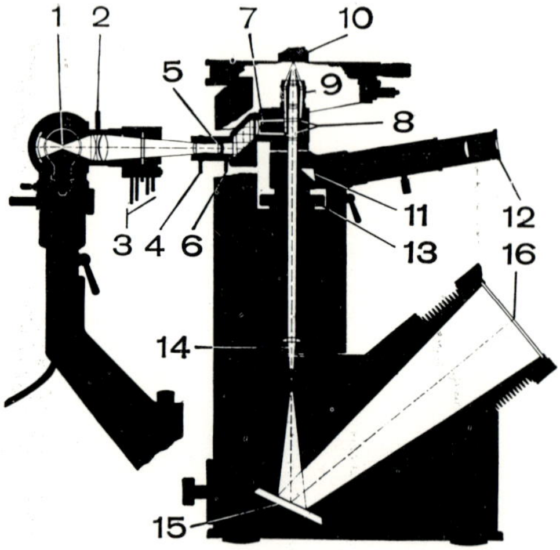

FIG. 3(b). Light path through the microscope shown in (a). 1 lamp filament. 2 condenser. 3 filter. 4 aperture-iris diaphragm. 5 deviating system. 6 field-iris diaphragm. 7 central diaphragm. 8 diverting mirror of opaque illuminator. 9 objective. 10 specimen. 11 diverting prism (swung out). 12 visual eyepiece tube (not used). 13 camera shutter. 14 photographic eyepiece. 15 diverting mirror of camera. 16 ground glass screen or photographic material (plate or film).

The most important component of the microscope is the objective, the chief function of which is to collect the maximum quantity of light coming from any part of the object. The numerical aperture (denoted by N.A.) of the objective is a measure of the light gathering power and is defined as

$$\text{N.A.} = \mu \sin \theta,$$

where μ is the refractive index of the medium in front of the objective and θ is half the angle of the aperture, i.e., half the angle subtended by the objective at the object, as shown in Fig. 4. It is evident from the above equation that the numerical aperture may be increased by replacing the air in front of the objective with a medium of greater refractive index, e.g., cedar oil. Typical specifications of several objectives are listed in Table 1.

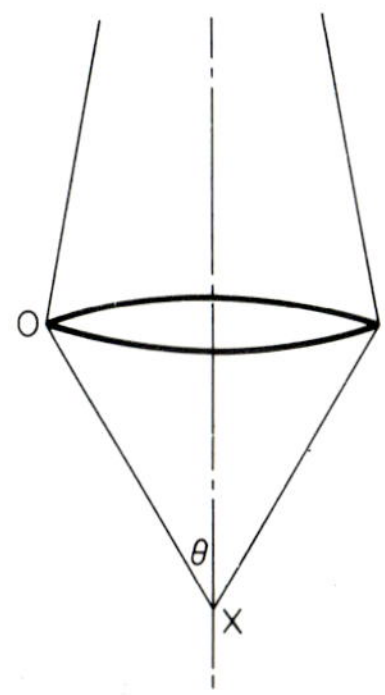

FIG. 4. Illustrating the angle θ which the objective O subtends at the specimen X.

TABLE 1. SPECIFICATIONS OF TYPICAL OBJECTIVES
(Taken from Cooke, Troughton & Simms, Ltd., publication CM 500L, p. 15)

Focal length (mm)	N.A.	Magnification	Working distance (mm)
33	0·10	× 6	38
25	0·15	× 9	16
16	0·28	× 16	5
8	0·50	× 28	1·7
4	0·65	—	—
4	0·85	× 60	0·46
1·8	1·30	× 135	0·35

The resolving power of an objective is defined as its ability to reveal closely adjacent structural details as actually separate and distinct. Quantitatively, this is given by the relation

$$\text{Resolving Power} = \frac{2 \times \text{N.A.}}{\lambda},$$

where λ is the wavelength of the light used. A quantity more in use than the resolving power is the limit of resolution, and this is the maximum distance between such details when resolution has been achieved. In quantitative terms, this is given by

$$\text{Limit of Resolution} = \frac{1}{\text{Resolving Power}} = \frac{\lambda}{2 \times \text{N.A.}}$$

It should now be evident why the numerical aperture is the most important property of the objective, since it is the numerical aperture and not the magnification which is a measure of the usefulness of an objective. Once the limit of resolution (which is inversely proportional to N.A.) has been reached, there is no point in increasing the magnification (e.g., by taking a photographic plate and enlarging it several times) since no extra detail will be revealed.

The magnifying power of a compound microscope can be shown to be

$$M = \left(1 - \frac{v}{f_o}\right) \left(1 + \frac{D}{f_e}\right)^{*},$$

where v is the distance from the objective to image I_1 in Fig. 1, D is the least distance of distinct vision (~ 25 cm for most people), and f_o and f_e are the focal lengths of the objective and eyepiece respectively. Manufacturers of microscopes usually call $\left(1 - \frac{v}{f_o}\right)$ and $\left(1 + \frac{D}{f_e}\right)$ the magnifying powers of the objective and eyepiece respectively, so that numerical guides to magnification can be stamped on objectives and eyepieces. Typical

*This formula was derived using the optical convention that real is positive.

values for objective magnifying powers are given in Table 1. An approximate value for the total magnification of the object under examination is then obtained by multiplying the magnifying powers of the objective and eyepiece. However, since the numerical values of v and D depend on the observer, an accurate measurement of the magnification can be made only by measuring the magnification of a calibrated grating. A calibrated grating is a series of fine, equi-spaced, parallel lines ruled on a polished metal surface; the width of the grating is equal to $(a+b)$, where a is the width of the (clear) opening and b is that of the (opaque) line, and is known beforehand. Measurement of the magnified grating width is made either by counting the number of lines in a known field of view, or by photographing the grating and measuring the width on the photographic plate. The magnified grating width divided by the true grating width gives the magnification.

Another important property of the objective is its depth of focus. This is the property which determines how much of the third dimension of the specimen—that of depth—is sharp and well defined at the same setting of the microscope. It can be shown that the depth of focus is proportional to $1/(\text{N.A.})^2$. Thus, for all the "hills and valleys" of an uneven surface to be in focus, we must use an objective with small N.A.; such an objective has a small magnification.

DEFECTS OF LENSES

Ordinary lenses do not give perfect images; various defects or aberrations occur. There are five aberrations with monochromatic light, namely

1. Spherical aberration
2. Coma
3. Astigmatism
4. Curvature of field
5. Distortion

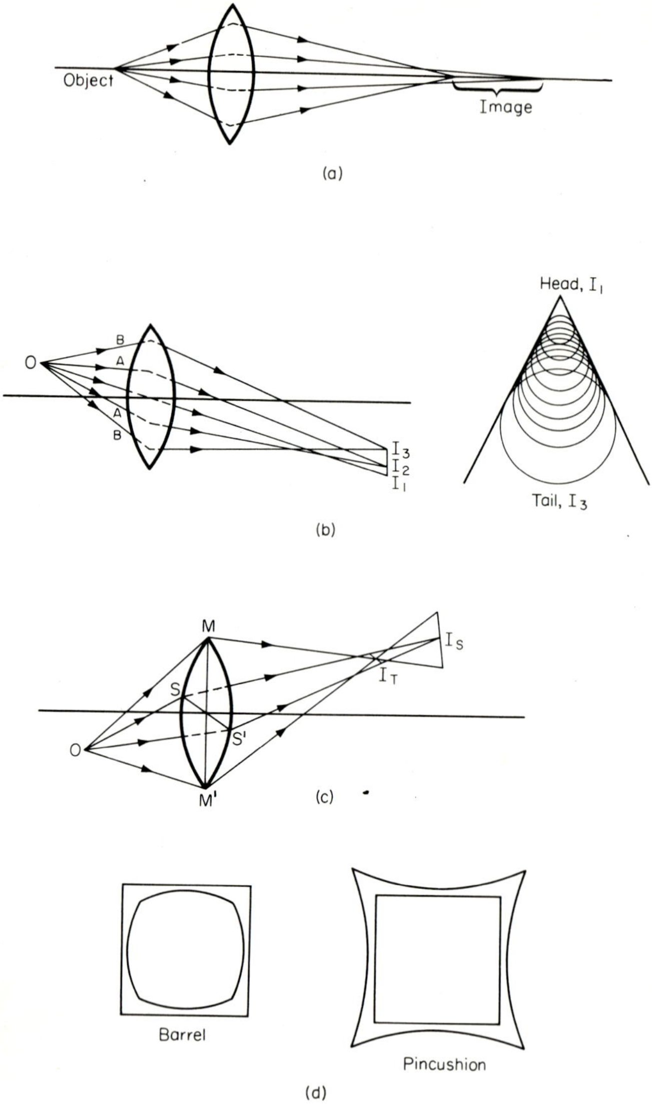

FIG. 5. Defects of lenses: (a) spherical aberration, (b) coma, (c) astigmatism, (d) distortion.

When white light is used, two additional aberrations appear:

6. Longitudinal chromatic aberration
7. Transverse chromatic aberration
   (chromatic variation of magnification)

These defects preclude the use of simple lenses in optical instruments and will now be briefly discussed in order to show how they may be removed or reduced to tolerable magnitudes.

Spherical aberration manifests itself as a longitudinal difference in focus between rays coming from different regions of the lens. In Fig. 5(a), rays travelling near the principal axis of the lens form an image on the axis at points further from the lens than do rays further from the axis. This is called positive spherical aberration. In cases where the innermost rays produce foci nearer to the lens than the outermost ones, the aberration is negative. A convenient method of reducing spherical aberration is to place a diaphragm in front of the lens so that only rays close to the principal axis go on to form the image. The main drawback of such a diaphragm is that it reduces the brightness of the image because some light is lost. An alternative solution, which is used in the design of oil-immersion (high magnification) objectives, is to position the object at an aplanatic point. The image formed is then free from spherical aberration (and coma) without the necessity of a diaphragm, i.e., without any reduction in its brightness.

Even though a lens may be corrected for spherical aberration for objects situated on the principal axis, spherical aberration may still occur for objects placed off the axis. This defect is known as coma and is illustrated in Fig. 5(b). The principal ray from the non-axial object O produces image I_1. The rays labelled A pass through the lens in a zone of greater radius and form image I_2, while those marked B pass through an outer zone to form image I_3. Consequently, the image of the point object O is spread over the length I_1 I_3. In addition, the complete ring of rays A from the whole of that zone of the lens produces a circular image at I_2. The radius of these images increases with the radius of the zone involved. The complete image thus consists of a series of overlapping circular images of radius which increases from I_1 outwards. The greatest intensity of light will be at I_1, the image becoming progressively fainter as we move away from I_1. This forms a comet- or coma-like image as illustrated in Fig. 5(b). It is evident that coma arises because different parts of a lens produce different linear magnifications of the same object. The effect is small when only the inner regions of the lens are used, i.e., when the outer regions are masked by an aperture.

The image of a point object located a long way from the principal axis is characterized by two focal lines with a circle of least confusion between them. This defect, known as astigmatism, is illustrated in Fig. 5(c). The tangential* rays OM and OM' produce the tangential image I_T, and the sagittal rays form the sagittal image I_S. In fact, a horizontal line image is formed at I_T and a line image lying in the tangential plane at right angles to the principal ray at I_S. The circle of least confusion lies between them and is the best image obtainable.

The defect known as curvature of field causes the image of a flat object to be formed on a curved instead of a flat surface. This curvature is due to the change in focal length of the lens as the position of a point object moves away from the principal axis. Consequently, this defect depends upon the refractive index and radii of curvature of the lens and can be avoided by careful design.

Distortion arises from the variation of magnification with the distance of the object from the optical axis. If the magnification decreases with increasing distance from the axis, the outer regions of the image are on too small a scale and barrel distortion results. If, on the other hand, the magnification increases as we proceed outwards from the optical axis, the outer parts of the image are on too large a scale thus producing pin-cushion distortion. Both types of distortion are illustrated

*The vertical diameter MM' of the lens and the horizontal diameter SS' define the tangential and sagittal planes respectively.

in Fig. 5(d). Distortion occurs when the principal ray from an off-axis object is prevented (by a lens stop) from entering a lens. Thus careful positioning of stops is required in order to prevent distortion.

All glasses exhibit the property of dispersion; the refractive index is not the same for light of different wavelengths. Consequently, the image formed by a convex lens from rays with long wavelength is further from the lens than that formed by rays with shorter wavelength. Thus, the image of a point object illuminated with white light is a blue spot surrounded by a red annulus as shown in Fig. 6. This defect is called longitudinal

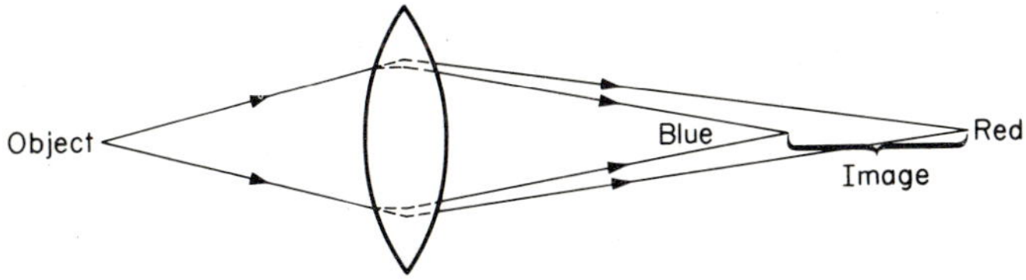

FIG. 6. Ray diagram illustrating chromatic aberration.

chromatic aberration. It can be removed by combining a convex lens of crown glass with a weaker concave lens of flint glass, thus forming an achromatic doublet.

Transverse chromatic aberration causes off-axis point objects illuminated with white light to produce images in the form of small linear spectra oriented radially with respect to the centre of the field. One result of this defect is that the magnification varies with colour. Transverse chromatic aberration can be removed by having suitably designed combination lenses.

METHODS OF INCREASING THE RESOLVING POWER OF AN OBJECTIVE

The numerical aperture, and hence the resolving power, of an objective is fixed by the geometrical optics of the system (N.A. $= \mu \sin \theta$, see previous section). The only variables are the refractive index of the medium in front of the objective and the wavelength of the light used. Under ordinary laboratory conditions, the largest N.A. is obtained using an oil ($\mu \sim 1{\cdot}3$) immersion objective in conjunction with green light ($\lambda \sim 5000$ Å). The theoretical limit of resolution is then $\sim 0{\cdot}2$ microns, but in practice this limit is more nearly 1 micron. Oils with refractive indices greater than $\mu \sim 1{\cdot}3$ are not compatible with the necessary property of optical transparency, and hence cannot be used. As a consequence, the only variable left in the expression governing the resolving power is the wavelength of the light to be used.

The wavelength of ultra-violet light is ~ 2000 Å and can therefore be used to decrease the limit of resolution to about half the value for green light. It follows that if one is close to the limit of resolution when using green light, it may be advantageous to change the radiation from green to ultra-violet light. A practical difficulty encountered when using ultra-violet light is that an objective made of quartz or fluorite must be used to transmit this wavelength, since ultra-violet light is heavily absorbed by normal glass lenses.

Later chapters will deal with electron microscopy for which a substantial gain in resolving power is obtained by virtue of the very small wavelength associated with a beam of electrons, e.g., $\lambda = 0{\cdot}037$ Å for an accelerating voltage of 100 kV.

EXAMPLES OF THE CONTRAST OBSERVED WITH REFLECTED LIGHT

Features of metallurgical specimens such as grains, grain boundaries, cracks and precipitate particles can be distinguished from each other because they differ in contrast. The nature of the contrast observed in ordinary reflected light microscopy arises from differences in the amount of light reflected into the objective by the various features in the specimen examined. These differences are caused by differences in (a) the amount of light absorbed and (b) the inclination of those atomic planes which reflect the light.

Examples of contrast due to differences in absorption commonly occur in everyday life. For example, objects viewed by white light are distinguishable one from another purely because of differences in the wavelengths which each object absorbs. Similarly the particles of graphite in the polished section of a cast iron, shown in Fig. 7, appear dark compared with the

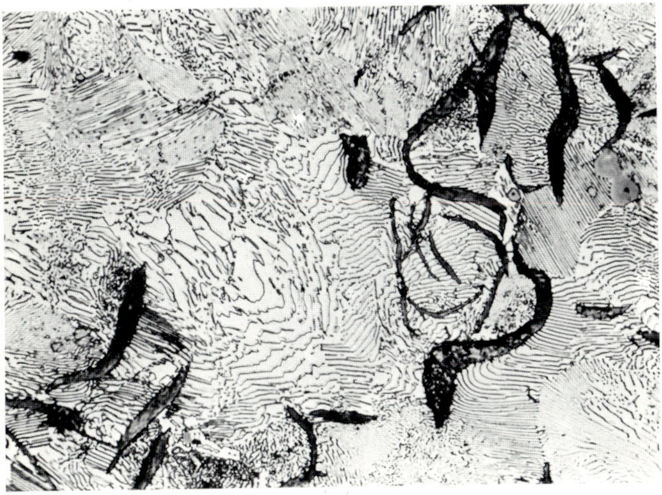

FIG. 7. Cast iron unetched. Magnification ×280. The particles of graphite show dark contrast owing to high absorption. (*Courtesy of Leitz.*)

surrounding matrix because they absorb more (and therefore reflect less) wavelengths than the latter. Contrast due to absorption is also responsible for the appearance of cored structures found in alloys cooled rather rapidly in casting. In this case, differences in staining by the etchant occur because of the composition variations.

A typical example of contrast arising from differences in the inclination of the planes of atoms reflecting the incident rays of light is the various shades of darkness in differently oriented grains of an etched polycrystalline specimen. Suppose, for

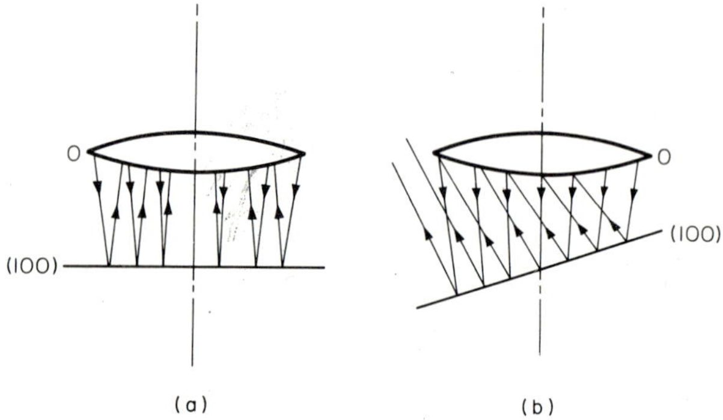

FIG. 8. Contrast due to differences in orientation. A grain with orientation (a) reflects more light into the objective and appears brighter than a grain with orientation (b).

FIG. 9. The grain structure of annealed cold-drawn α-brass as revealed by differences in grain orientation. Magnification ×approx. 130. (From Greaves and Wrighton, *Practical Microscopical Examination of Metals,* fig. 57 D.)

example, that we have a cubic metal in which (100) planes are preferentially attacked by the etchant. The quantity of light reflected into the objective will depend on the orientation of the (100) planes of different grains with respect to the incident rays as shown in Figs. 8(a) and (b). Figure 9 shows an example of the variation of contrast as the orientation changes from grain to grain in a polycrystal.

It should be noted that a variation in orientation from grain to grain does not always produce a corresponding variation in contrast. An example of this given in Fig. 10 is the grain

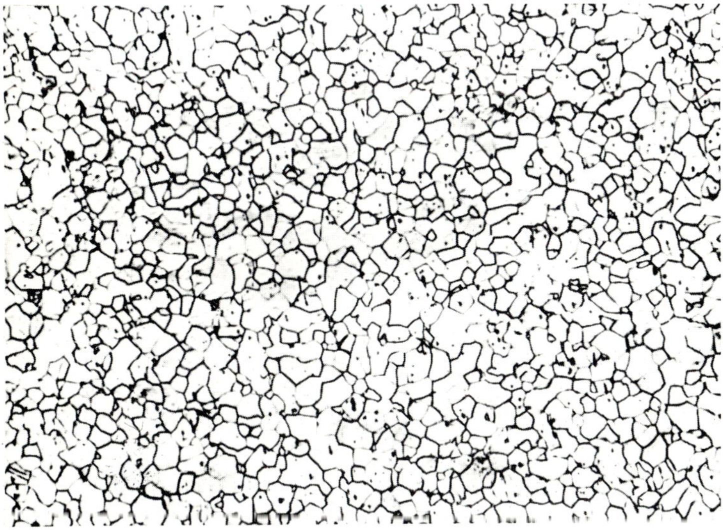

FIG. 10. The grain structure of Armco iron revealed by preferential attack at grain boundaries and not by orientation variations from grain to grain. Magnification ×28. (*Courtesy of Leitz.*)

structure observed in Armco iron. One explanation of the the absence of contrast even when crystals change in orientation is that the etchant has attacked the surface uniformly and not preferentially along certain crystallographic directions. The surface being examined therefore remains flat, and is everywhere normal to the incident rays so that the objective gathers equal amounts of reflected light from all the grains. The contrast at the grain boundaries in Fig. 10 is due both to staining and preferential dissolution during etching.

SUGGESTIONS FOR FURTHER READING

R. G. Greaves and H. Wrighton, *Practical Microscopical Metallography*, Chapman & Hall, 1957.

L. C. Martin and B. K. Johnson, *Practical Microscopy*, Blackie, 1949.

CHAPTER II

HIGH TEMPERATURE MICROSCOPY

INTRODUCTION

With the ordinary light microscope it is possible to study the structure of as-received specimens, as well as the effect of heat treatment, e.g., annealing, normalizing and quenching of steels, on microstructure. However, if a microscope could be used only at room temperature, the scope of metallography would be considerably restricted. Thus, it would be impossible to study those alloys containing phases which are stable only at high temperatures. Such alloys undergo a transformation on cooling and, as a consequence, only the decomposition products are available for study at room temperature. It would also be impossible to examine the kinetics of such phenomena as recrystallization, grain growth, polymorphic transformations and graphitization.

It is evident that a real need exists for high temperature microscopy. One of the difficulties encountered when studying microstructures at elevated temperatures is oxidation and to avoid this, provision is usually made for heating the specimen either in a vacuum or in an inert atmosphere. The specimen chamber is fitted with an optically-worked silica window through which the specimen may be observed. Several methods of heating have been used including radiation, self-resistance, electron-bombardment and conventional furnace methods. Figure 11 shows a hot-stage microscope manufactured by Reichert for use at temperatures up to 1600°C in a vacuum of 10^{-6} mm Hg.

reach the objective) necessitates carefully designed illuminating devices for bright field examination of opaque specimens. In consequence, the attachment is used only for studying transparent specimens with transmitted light and opaque specimens with reflected light by dark-field illumination.

Burch has built a completely reflecting objective for use in a specially designed microscope. The advantages of a fully reflecting objective are perfect achromatism, absence of absorption in the ultra-violet and infra-red regions of the spectrum and an extremely large working distance. The basic components of this type of objective, shown in Fig. 13, are a

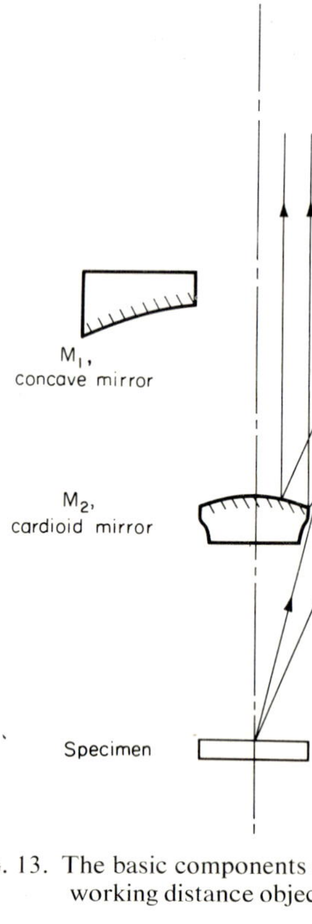

FIG. 13. The basic components of Burch's long working distance objective.

concave mirror M_1 and a small cardioid mirror M_2 supported between M_1 and the specimen. An objective of this design has a numerical aperture of 0·65 (comparable to that of a 4 mm normal objective) and a working distance of 14 mm.

TYPICAL HOT-STAGE EXPERIMENTS

It is well known that when a deformed metal is annealed, new or recrystallized grains grow at the expense of the old deformed ones. The growth of these new grains is best studied metallographically at the recrystallization temperature using a hot-stage microscope. Figure 14 shows a sequence of grain growth photographs in electrolytic copper. The contrast at the grain boundaries is produced by thermal etching, and hence the absence of grain boundary contrast in the "as deformed" state shown in Fig. 14(a) throughout the sequence. The recrystallized grains first appear at about 300°C and these new grains grow in size as the temperature is raised.

An estimate of the grain sizes can be made using a special eyepiece inserted into the microscope in place of the normal eyepiece. The grain size is estimated by comparing the grains under observation at a known magnification, with shapes corresponding to some standard which are reproduced on a graticule built into the image plane of the eyepiece. Various standard shapes of networks are reproduced in Fig. 15. Using such graticules, grain size measurements can be made as a function of both temperature and time to give valuable data regarding the kinetics of grain growth. For example, measurements reproduced in Fig. 16 have shown that austenitic grain growth proceeds by at least four different mechanisms. Curve 1 is typical of monotonic grain growth, but curves 2, 3 and 4 are all characterized by sudden rapid increases in grain dimensions at critical temperatures. It is evident that small temperature increases near these critical temperatures would lead to rapid grain growth which, with its associated changes in mechanical properties, may be undesirable.

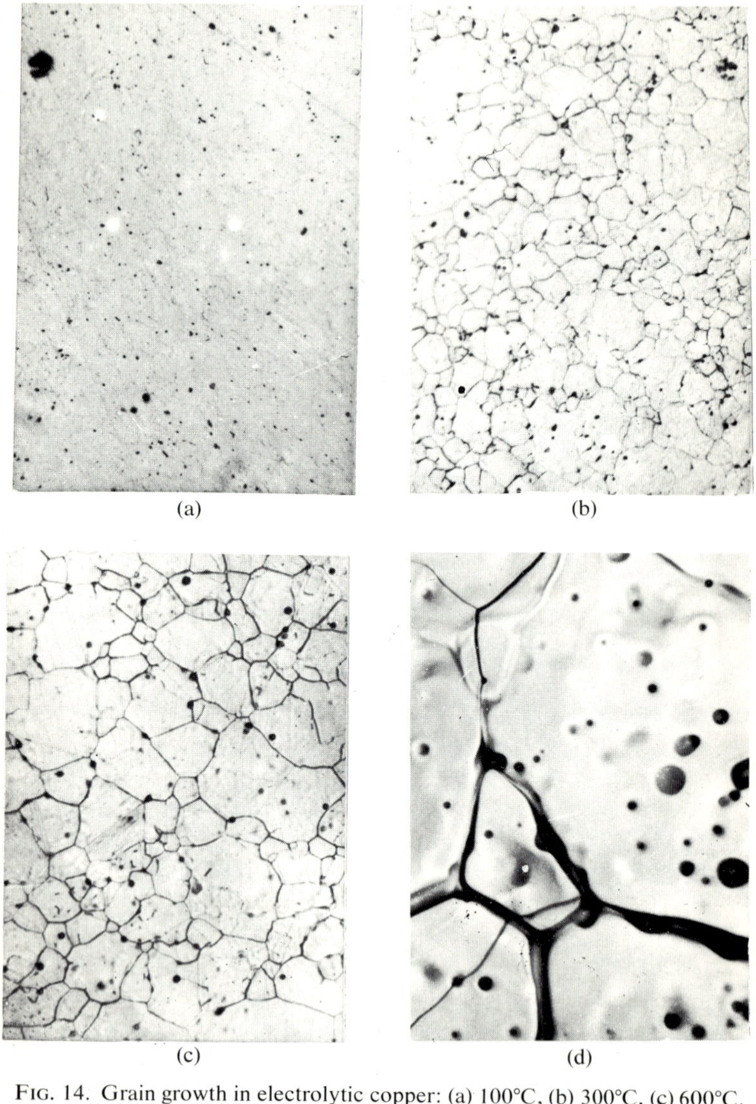

(a) (b) (c) (d)

FIG. 14. Grain growth in electrolytic copper: (a) 100°C, (b) 300°C, (c) 600°C, (d) 1050°C. (After F. Gabler and W. Wurtz, *Metall.*, **9,** 819–823, 1959.)

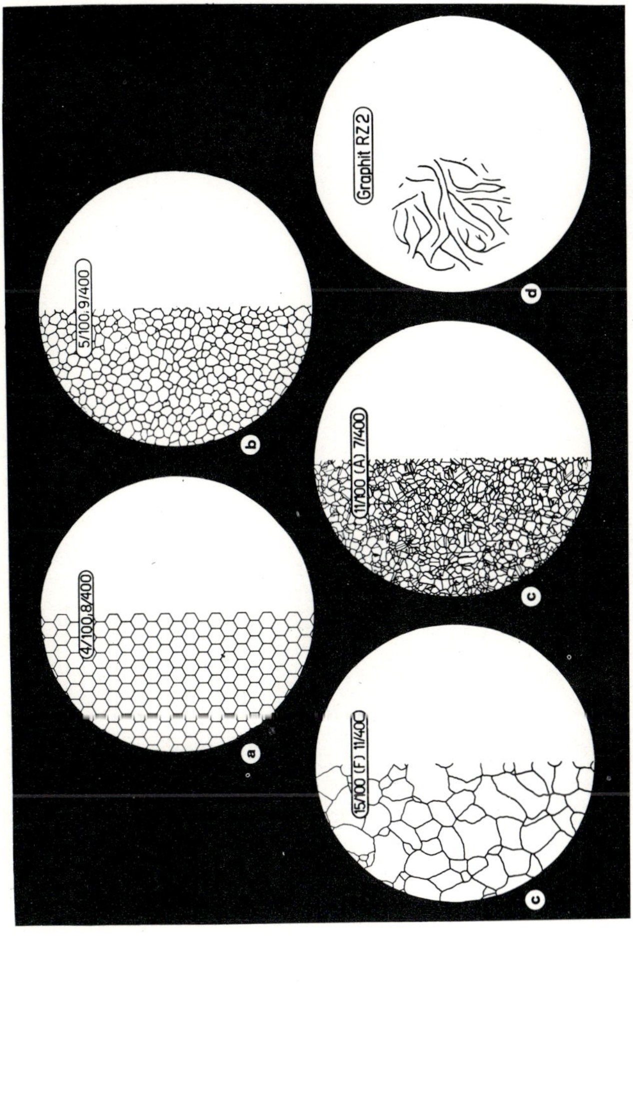

FIG. 15. Grain size networks (a) ASTM hexagon, (b) ASTM irregular, (c) ferrite and austenite, and (d) graphite screen. (After F. Gabler, *Mikroscopic* **11,** No. 7/8, 2, 1956.)

There are many such experiments which can be carried out using a hot-stage microscope, and these include studies of

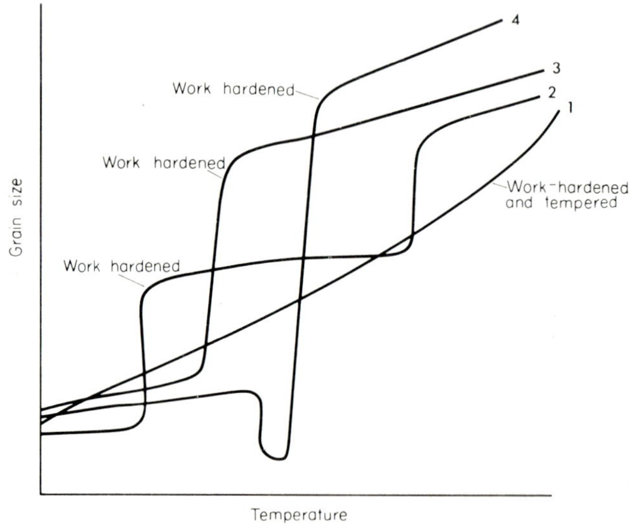

FIG. 16. Grain growth of austenite. (After M.G. Lozhinskii, *Metallovedeniye*, No. 11, 18–42, 1957.)

phase transformations and precipitation. Figure 17 shows a series of micrographs taken during the formation of bainite in a steel containing 0·45%C, 2%W and 1%Cr. Micrograph (a) shows the relatively fine alpha grain structure observed at 950°C as a network of bright lines. The coarse gamma grain structure, shown up through thermal etching, is also visible. The specimen was then cooled to 450°C which corresponds to the bay of the T.T.T. curve, and held for a total of 5 minutes during which time, micrographs (b), (c) and (d) were taken. These photographs show that bainite formation is by no means instantaneous.

Before closing this chapter, a further important use of high temperature microscopy should be mentioned. This is the

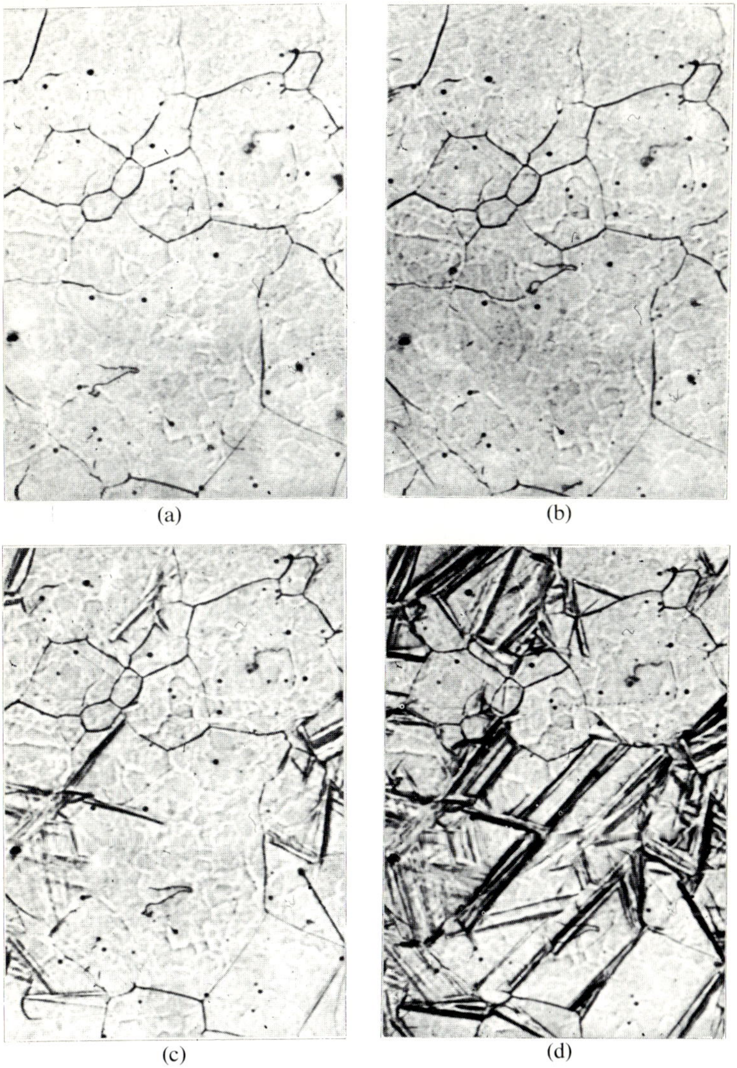

FIG. 17. Formation of bainite in a steel containing 0·45%C, 2%W, 1% Cr. (After R. Mitsche, F. Gabler and W. Wurtz, *Aluminium,* **37,** No. 10, 652, 1961.)

metallographic observation of mechanical properties, such as hardness and tensile behaviour, as a function of temperature. Special hardness indentors and straining jigs are commercially available for many microscopes and are useful for studying and assessing the mechanical properties of metals and alloys at elevated temperatures.

SUGGESTIONS FOR FURTHER READING

J. DYSON, *Proc. Phys. Soc. Lond.*, B **62**, 565 (1949).
C. R. BURCH, *Proc. Phys. Soc.*, **59**, 41 (1947).
M. G. LOZINSKII, *High Temperature Metallography*, Pergamon Press, 1961.

CHAPTER III

SURFACE TOPOGRAPHY

THE IMPORTANCE OF SURFACE TOPOGRAPHY OBSERVATIONS

The study of surface features on metallographic specimens is known as surface topography. Surface relief effects may arise either from differential attack during polishing and etching, or from mechanical deformation which may lead to slip lines and twin boundaries within the grains and cleavage steps on fracture surfaces. In industrial practice, examples of surface relief and surface variations often occur as a result of wear and exposure of metal components. The metallographic examination of such topographical features can often provide considerable information regarding the mechanisms which produce them, and hence in this chapter an account is given of the various methods of studying and measuring such surface features.

Surface reliefs can be divided into two categories, namely steps and tilts, which are illustrated respectively in Fig. 18 (a) and (b). Quantitative topographic studies concern the measurement of both step heights and angles of tilt. With a normal high resolution microscope, step-heights of about 100 Å can be detected by the faint change in contrast caused by diffraction scattering at the step, but the step-height cannot be measured. Tilts examined with a normal microscope can be seen only if the inclination is so large that the light reflected from the tilt is not collected by the objective as illustrated in Fig. 18(c). The angular sensitivity of a conventional microscope to a surface tilt is inversely proportional to the sine of the numerical aperture, and for a given objective can be improved only by offsetting the light source, i.e., by having non-central illumination as shown by the spaced rays in Fig. 18(c). With such illumination, however, the resolving power is lowered because the objective is not completely filled with light. A useful

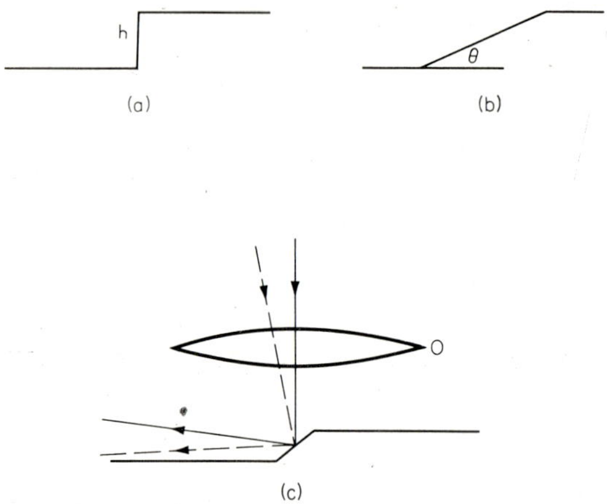

FIG. 18. Surface topography is divided into (a) steps measured by the step-height h, and (b) tilts measured by the angle of tilt θ. (c) Light reflected by large tilts misses the objective and the tilt appears in dark contrast.

method of detecting tilts with a normal microscope is to use an oil immersion objective when the tilt can be seen as a change in height; the high magnification of an oil immersion objective means a small depth of focus which makes it impossible to have the top and bottom of the tilt in focus simultaneously.

In general, the methods of studying surface topography can be divided into three groups, namely oblique illumination, modifications of the normal microscope and specific surface contouring techniques. Each of these groups will now be considered in detail.

OBLIQUE ILLUMINATION

The use of non-central illumination produced in an ordinary microscope by offsetting the light source was mentioned in a previous section. Oblique illumination can also be produced by moving the normal light source to the outside of the microscope as shown in Fig. 19. Usually it is difficult to get sufficient light in under the objective and hence a long working distance objective may be required for this technique. Combined with a rotating stage, this method can provide a good indication of the surface relief but cannot be used to measure surface contours.

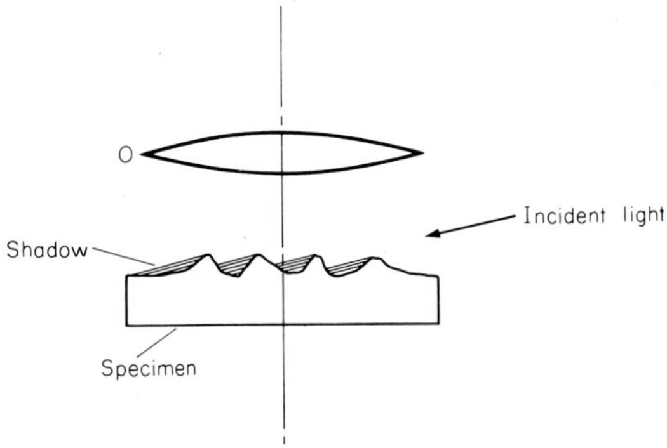

FIG. 19. Oblique illumination with a normal microscope.

Surface relief can be measured by obliquely shadowing the surface with a heavy metal and observing the shadows using normal illumination. The apparatus, schematically shown in Fig. 20(a), consists of a vacuum coating unit. The heavy metal, usually silver or gold in the form of a bead, is contained together with the specimen inside the bell-jar and is evaporated by electrical heating. The heavy metal atoms strike the surface to be examined at a low and fairly constant angle θ (see Fig. 20a) and the surface undulations are coated with a heavy metal shadow on the faces remote to the evaporated bead, as shown in Fig. 20(b). Knowing the angle θ and by measuring l, the length of a shadow, step-heights as small as about 50 Å can be calculated.

Another variation of oblique illumination is the light profile method, the principle of which will be illustrated by the following analogy. Consider an observer standing at an angle of 45° to a flight of steps with a lamp held in his hand so as to cast a shadow of a vertical stick upon the steps. The view he sees is

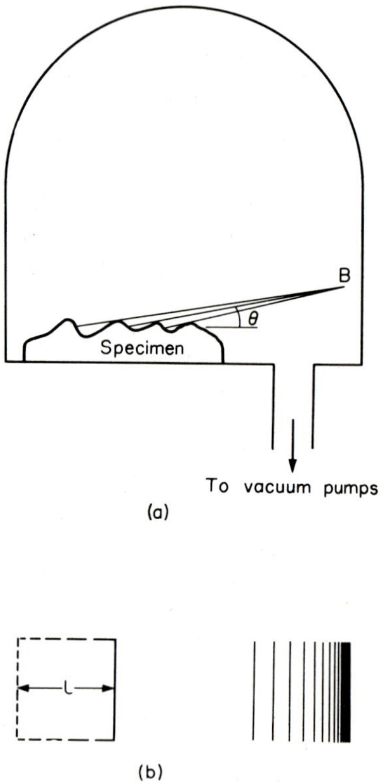

FIG. 20. (a) Apparatus for shadowing surface features with a heavy metal. (b) The appearances of a surface after shadowing.

shown in Fig. 21(a). A second observer standing directly in front of the steps sees a different view of the shadow, and this is shown in Fig. 21(b). From the second observer's position, the effect is to convert a profile in depth into a line pattern in extension and, by measuring the offset *xy* in Fig. 21(b), he can obtain the depth of the step.

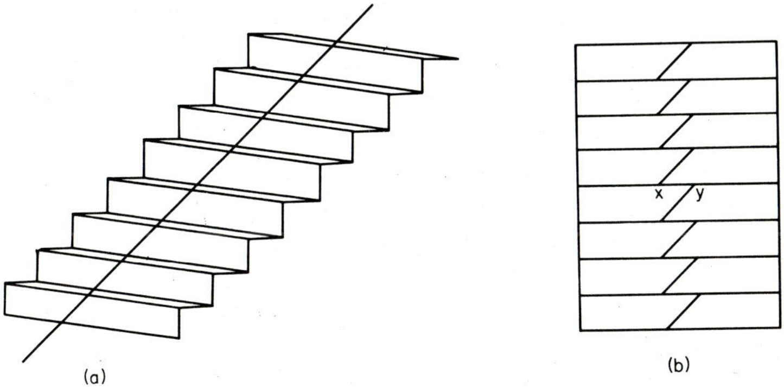

FIG. 21. Principle of the light profile technique.

For high resolution and high magnification, the light profile technique imposes four conditions: (1) a line image must be formed on the specimen surface, (2) the illumination forming this image must be oblique, (3) an oil immersion objective has to be employed and (4) most of the lens aperture should function. The Vickers Projection Microscope adequately satisfies these conditions if (1) a piece of wire is placed close to the field iris to produce the line image and (2) the normal 45° glass reflector is replaced with a metal-tongue sector so as to cast the line shadow and illuminate the surface simultaneously. Monochromatic (usually green) light must be used and the profile magnification is

$$M^1 = \frac{2M}{\mu} . \tan i,$$

where M is the usual microscope magnification in extension, μ is the refractive index of the immersing medium, and i is the effective angle of incidence as determined by the objective aperture and the position of the reflector. Rather than determine i, it is more convenient to calibrate the system by evaluating a step graticule both by interferometry (see p. 40) and light profile to get the ratio M^1/M. For example, a 2·2 mm apochromatic lens gives $M^1/M = 1$ which means that $i \sim 45°$. The method permits full use of the high resolving power and magnification (× 1200) of the oil immersion lens, and step heights as small as 1000 Å can be measured.

OPAQUE STOP AND PHASE CONTRAST

The techniques described in this section may be carried out by simple modifications of the normal microscope. Fig. 22

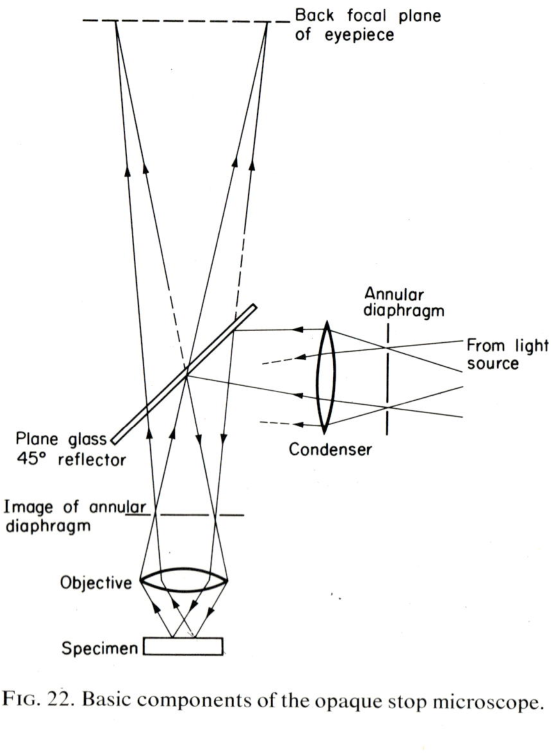

FIG. 22. Basic components of the opaque stop microscope.

shows a schematic ray diagram illustrating the principle of the opaque stop technique. A ring stop, in the form of a metal disc with an annular opening, is positioned between the light source and the condenser. By adjusting the condenser so that an image of the filament of the light source is formed at the ring stop, a hollow cylinder of light is brought to a focus at the specimen. When the microscope operates without the stop a perfectly flat specimen surface will reflect the light to form an image in the back focal plane of the objective (see Fig. 22) and thus if the opaque stop is then placed over this image, the eye will see nothing. Suppose, now, that the specimen surface consists of two areas inclined to each other by a small angle θ. Each area will produce its own image of the ring stop in the back focal plane of the objective and, referring to Fig. 23, the two

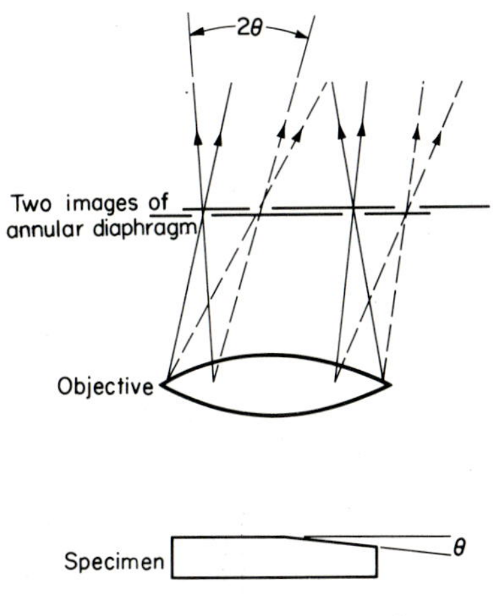

FIG. 23. Displacement of the image of the ring stop by a surface tilt.

images will be angularly displaced with respect to each other by 2θ. The image produced by the horizontal area is obscured by the opaque stop, but that formed by the inclined area will miss the stop and hence will be observed.

The opaque stop technique is a sensitive dark-field technique; any tilt on the surface is revealed as a bright image on a dark background. In addition to large surface contours, fine surface detail can be examined since it leads to random scattering and is observed as bright contrast on a dark background.

The technique of phase contrast microscopy permits those structural details to be observed which would not normally be distinguishable with an ordinary microscope, not because they are too small or too close together but because they do not give rise to any contrast. The technique depends on the fact that when a specimen diffracts the incident light, the diffracted light is out of phase with the directly reflected light by one quarter of a wavelength. In metallography, diffraction

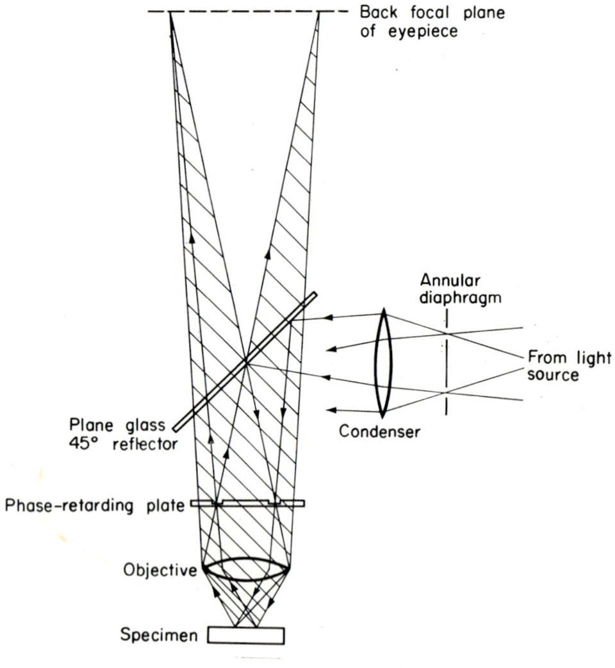

FIG. 24. Phase contrast microscope.

occurs from small surface reliefs and phase contrast provides an excellent technique for examining such features. Using a normal microscope, the difference between the diffracted and direct rays is not observed because the eye responds not to variations in phase but to variations in intensity. The function of the phase contrast microscope is to change differences in phase into intensity variations so that they can be detected by the eye.

Figure 24 shows a schematic arrangement of a phase contrast microscope. A metal disc or diaphragm with an annular hole cut in it, is placed at the front focal plane of the condenser (between the light source and the condenser) so that an image of the annular diaphragm is formed in the back focal plane of the objective, the location of the phase ring (see below). To describe the principle of phase contrast it is convenient to represent the light, which travels by means of vibrations, by sine waves, as shown in Fig. 25(a). The vibrations of the light diffracted by the surface are shifted slightly with respect to those of the direct light as shown by curves 1 and 2 respectively in Fig. 25(b). The two sine curves are identical in that they have the same amplitude and wavelength but there is a slight phase difference between them. Now curve 1 in Fig. 25(b) can be resolved into two components as shown in Fig. 25(c), one component coinciding with curve 2 and the other with a new curve which we shall denote 3. It is noticed that curve 3 is shifted by an amount $\lambda/4$, or $\pi/2$ in terms of phase angles, in relation to vibration 2 and has a much smaller amplitude, which is represented by OM, We can therefore consider that the image formed by the diffracted rays is the result of the superposition of two component vibrations which are out of phase by $\pi/2$. Since the intensity of a diffracted ray is proportional to the square of its amplitude, the intensity of of the diffracted image is proportional to the sum of the squares of amplitudes 2 and 3 and, since the summation must be made at the same value of phase angle for both curves (e.g., at $\pi/2$) the resultant intensity is simply the square of amplitude 2.

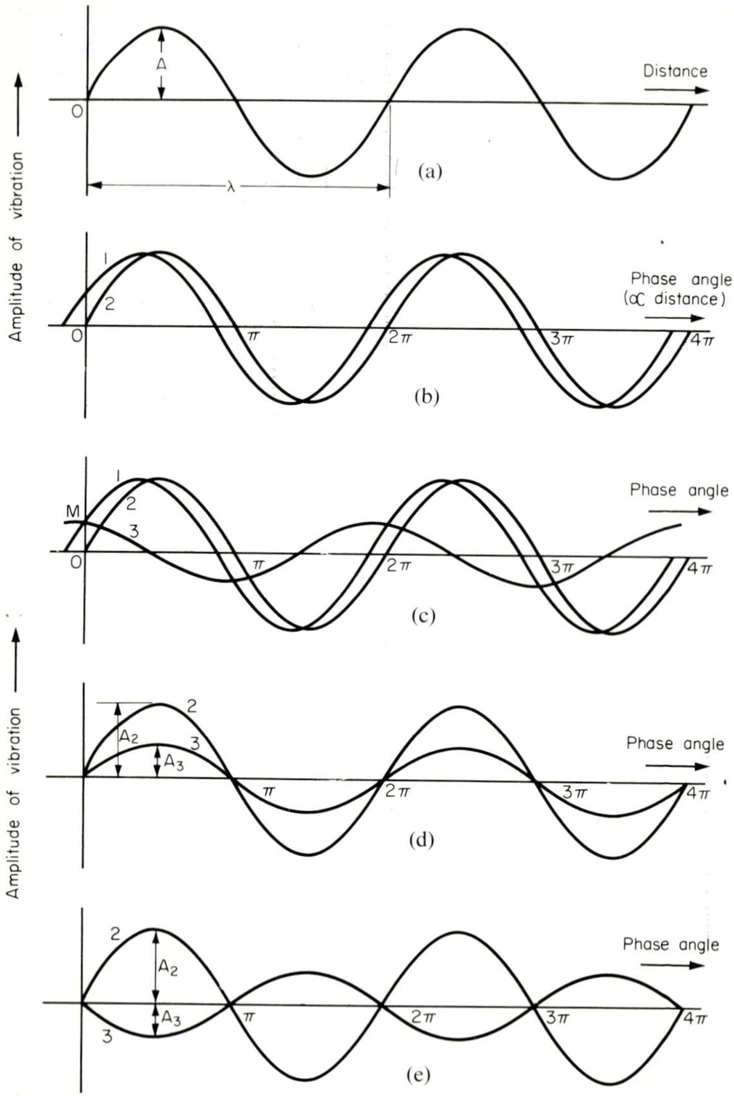

FIG. 25. Principle of phase contrast.

This intensity value is the same as that of the image formed by the directly reflected light, and so the two images are indistinguishable.

Now suppose that vibration 3 in Fig. 23(c) were to be shifted by $\lambda/4$ (i.e., $\pi/2$) in relation to vibration 2 as illustrated in Fig. 25(d) or (e) depending on whether the shift is in a forward or backward sense. The two vibrations are exactly in phase and the intensity of the diffracted image is proportional to the sum of the squares of the two amplitudes A_2 and A_3. This is greater than the square of amplitude A_1 alone, which is proportional to the intensity of the normal image formed by the directly reflected rays. The phase difference is therefore transformed into an intensity difference and hence the two images are now distinguishable one from the other. If, as in Fig. 25(e), curve 3 lags behind curve 2 by $\pi/2$, the intensity of the diffracted image is then proportional to the difference of the squares of amplitudes 2 and 3, and, since this is less than the intensity of the normal image, the images are again distinguishable.

With the phase contrast technique, surface features become visible because the diffracted vibrations are brought exactly into or out of phase with the direct vibrations. In practice, this phase shift is produced by placing a transparent phase-retarding ring of the same shape as the annulus of the condenser diaphragm in the back focal plane of the objective (see Fig. 24). The phase plate is usually made from glass and the phase changing properties of its annulus or ring are obtained either by etching away a ring shaped area or depositing a suitable thickness of transparent material.

Vibrations of the light passing through the image of the condenser diaphragm annulus and therefore through the phase ring are retarded, or advanced, by one quarter of a wavelength. Since the direct light is brought to a focus at the back focal plane of the objective, while the diffracted light is still diffuse, only the direct light is affected by the insertion of the phase plate. Two kinds of phase-contrast image are possible. For the positive type, the ring has a shorter optical path than the rest of the phase plate and for the negative kind the reverse occurs; a depression in the specimen would appear in dark contrast with positive-type phase plate.

The optical density of a given phase plate cannot provide ideal phase contrast conditions for all specimen surfaces. However, it is possible to increase the contrast by varying the relative amplitudes of the direct and diffracted beams by altering the thickness of the deposit on the phase plate. The relative amplitude can also be altered using polarized light (see Chapter 4), when the contrast is adjusted to suit the surface being studied simply by rotation of an analyser.

By its very nature, the phase contrast technique is mainly qualitative. The image of fine surface detail suffers from a fringe or "halo" effect because, unlike the direct light, the light diffracted from such detail does not pass through the phase ring. This optical artefact prevents really accurate measurements of surface features showing variations in step heights and consequently, although steps as small as 5 Å high are easily visible, the exact magnitude of the heights cannot be deduced.

A development in phase contrast microscopy is the interference contrast technique described by Françon and Nomanski. The main feature of this development is that polarized light is used to increase the sensitivity to extremely small variations in the height of the surface under examination. The sensitivity is adjusted, by rotating the analyser, to suit the surface being studied.

INTERFEROMETRY

During the last twenty years, specific surface contouring techniques have been developed which make use of optical interference fringe contours to show surface structure. The methods described below are simple in principle and may be used to give accurate optical contour maps, usually called interferograms, of the structure of the surface examined.

The simplest technique is that based on the interference between two beams of light. Referring to Fig. 26, monochromatic light from the source of illumination is divided by a beam splitter (an optical device which transmits and reflects equal proportions of the light it receives) into two equal beams at right angles to each other. One beam passes through a high power objective and is brought to a focus at the surface under examination, and the other beam, using an identical objective,

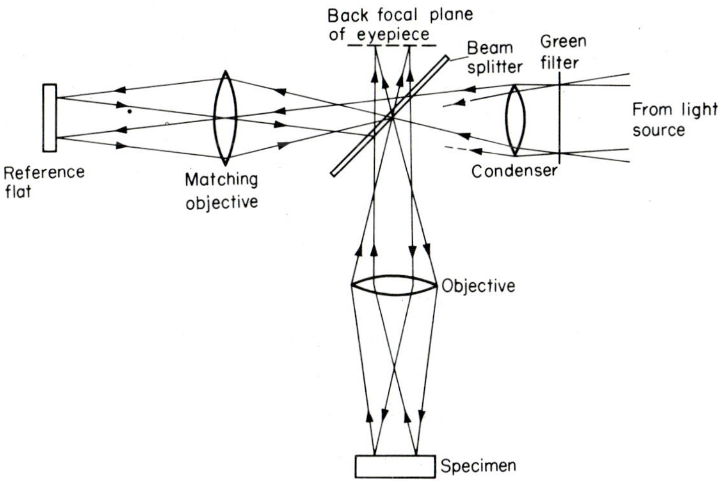

FIG. 26. The principle of two-beam interferometry.

is focused onto a metallized optical flat. The beam reflected from the irregular surface under test and that reflected by the flat standard surface meet at the beam splitter. These two beams reinforce each other at points where the optical path difference between them is equal to, or multiples of, one half the wavelength of the light used, and interfere with each other elsewhere. In this way optical contour lines or fringes are produced, and these contours make up an interferogram in which areas at the same level are joined either by a light line or a fringe.

In effect, the above technique matches the relief of the surface being studied against that of the reference flat. For maximum contrast, the two beams should have equal intensities and the reflectivity of the flat should be the same as that of the specimen surface. The change in height from one fringe to the next is $\lambda/2$, e.g. 2500 Å for the green-filtered line from a mercury arc light, and in practice displacements in contour lines can be estimated to about 1/10 of a fringe separation, i.e., to variations in height of about 250 Å. Apart from this high magnification in depth, the high resolution of the objective used permits high magnification (up to $\times 1200$) across the surface.

Dyson has designed a two-beam interference apparatus which is positioned between the specimen and the objective of a normal microscope. Figure 27 illustrates the principle of

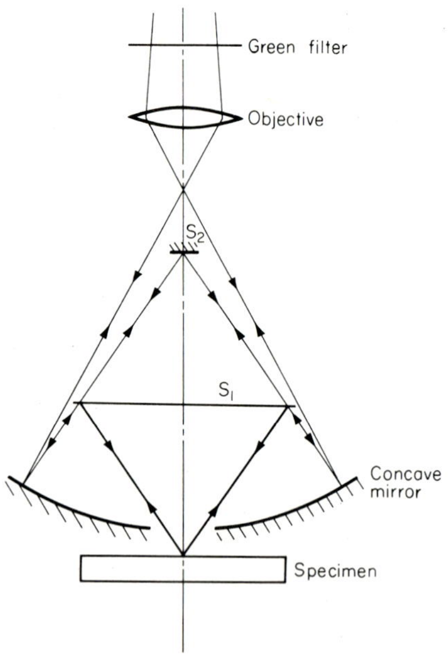

FIG. 27. Dyson's two-beam interference attachment.

this design. The light initially transmitted by the objective is divided at half-silvered surface S_1 such that part is focused on a silvered spot S_2 and part on the specimen surface. The two reflected beams interfere, as described above, at places where the optical path difference is not equal to, or multiples of, $\lambda/2$. This apparatus will resolve differences in level of 25 Å at a lateral magnification of ×1200. Since the reference flat S_2 is located below the main objective an advantage of this design is that an auxiliary identical objective is not required. The most important disadvantage is the delicacy of the apparatus which makes its accuracy susceptible to external influences such as vibrations and changes in ambient temperature.

A considerable increase of sensitivity is possible with multiple-beam interferometry where, instead of creating optical interference with two light beams, interference is produced between many (usually about 100) beams. The surface of the specimen and that of a reference flat are coated with a thin, partially transmitting but largely reflecting film of silver. The two surfaces are then brought into close parallelism (never more than one or two light wavelengths apart) and illuminated with a parallel beam of monochromatic light at perpendicular incidence, as shown in Fig. 28(a). There arrives at any point A a sequence of beams of diminishing intensity, each beam coming from a position, on the specimen surface, which is slightly displaced from that of the adjacent beam. The separation between the surfaces must be sufficiently small that the displacement, by multiple reflection, of even the hundredth beam is so small that all the beams can be regarded as coming from one very small area. The phases of successive beams gradually get further out of step with that of the first beam and interference results. An interference fringe is formed characteristic of the small area representing the multiply reflected beams, and the complete interference pattern accurately defines the microtopography of the area being studied.

The conventional metallurgical microscope can be modified as shown in Fig. 28(b) for multiple beam interferometry. The

45° glass reflector of the illuminating system is so positioned that an image of the light source is formed in the back focal plane of the objective. By this means, a parallel beam is directed

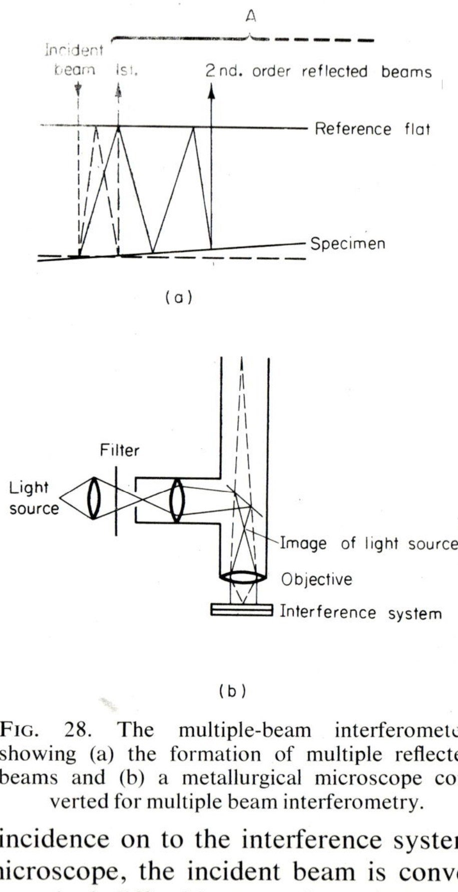

FIG. 28. The multiple-beam interferometer, showing (a) the formation of multiple reflected beams and (b) a metallurgical microscope converted for multiple beam interferometry.

at normal incidence on to the interference system (in an unmodified microscope, the incident beam is convergent). The two main practical difficulties are (1) accurate positioning of the glass reflector and (2) a long working distance objective is required because the reference flat (usually 2 mm to 3 mm thick) is placed above the specimen surface.

Compared with two-beam interferometry, the multiple-beam technique produces such refinement and sharpening in the fringes that the sensitivity is increased a hundredfold. Displacements of 1/500 of the fringe separation can be measured with confidence, and such a displacement represents a step height of the order of 5 Å. The versatility of the multiple-beam system permits accurate measurement of heights from 5 Å up to 1/100 mm.

There are four chief difficulties inherent in multiple-beam interferometry. The silver film must have a reflectivity of about 95%, at least 1% of the remainder being transmitted and the rest being absorbed. This requirement necessitates very careful preparation, purity and film thickness being critical. The second difficulty is that silver tarnishes very quickly, especially in industrial atmospheres, and the expected lifetime of the film is only an hour at the most. Thirdly, difficulties arise from the necessity of having to modify the illumination system to provide a strictly parallel beam of monochromatic light. The final limitation arises from the fact that a long working distance objective must be used, since the reference flat has to be positioned between the objective and the specimen, which implies a low lateral magnification of, say, ×200 (see Chapter I) and this restriction limits the apparatus to very small surface features which are fairly widely spaced.

A recent development in interferometry is the use of polarized, monochromatized light. The polarization interferometer is a simple accessory which is inserted between the illuminator and objective of a normal microscope, and consists essentially of a polarizer and a rotatable analyser. Interference fringes representing contour lines on the specimen surface are produced, in the "crossed polars" position, by interference between the two beams which form the double image in the double quartz prism. The two images are in fact slightly displaced and tilted relative to each other, so that each point on the specimen surface is compared with an adjacent point on the same surface, one image always acting as a reference flat for the other image. The use of the actual specimen surface as a reference flat in place of a comparison flat is one of the features of the polarization interferometer.

A completely flat surface produces interference fringes which are straight lines parallel and equidistant to each other. Any departure from true flatness results in a displacement of the fringes. This displacement is a measure of the depth irregularity producing it, a deviation by $1/n$ of a fringe spacing being equivalent to a step height of $\lambda/2n$ where λ is the wavelength of the light used. It is claimed that step heights in the range 10,000 Å to 100 Å can be measured by this method.

There is a serious limitation regarding magnification that is inherent in the principle of the polarization interferometer. Excessive magnification means that the displacement due to the double image is larger than the surface detail under examination. In other words, the surface irregularity must not be compared with parts of its own double image but only with that of the surrounding area.

EXAMPLES OF SURFACE TOPOGRAPHY

Two typical examples of the use of phase contrast are given in Fig. 29. The first micrograph, Fig. 29(a), shows an alloy steel structure after etching with alcoholic nitric acid. The etchant has preferentially attacked the specimen leaving some constituents in greater relief, and hence these constituents are distinguishable from others. Figure 29(b) shows the appearance of slip lines on an unetched surface of a ceramic crystal rutile, (TiO_2). The steps formed where active slip planes intersect the external surface are shown up by phase contrast as lines.

The advantage of interference contrast is illustrated by the examples given in Fig. 30(a) and (b) which show an etched section of cobalt steel viewed with normal light and interference contrast respectively. Small differences in level are often produced during polishing and these differences of height can be used to render such phases visible in high contrast. The three-dimensional appearance of electro-polished cobalt

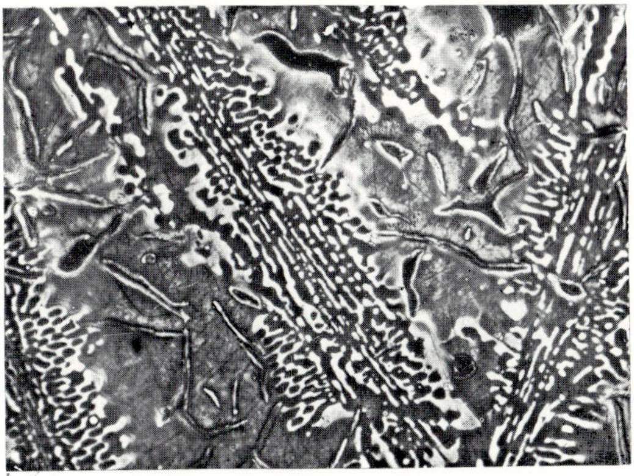

(a)

(b)

FIG. 29. (a) Alloy steel under phase contrast. Etchant: alcoholic nitric acid. Magnification ×540. (*Courtesy of Reichert.*) (b) Slip lines on the surface of rutile compressed 3% at 900°C. Magnification ×90. Phase contrast. (After Ashbee and Smallman, *Proc. Roy. Soc.* A, **274,** 195 (1963), fig. 2(b).)

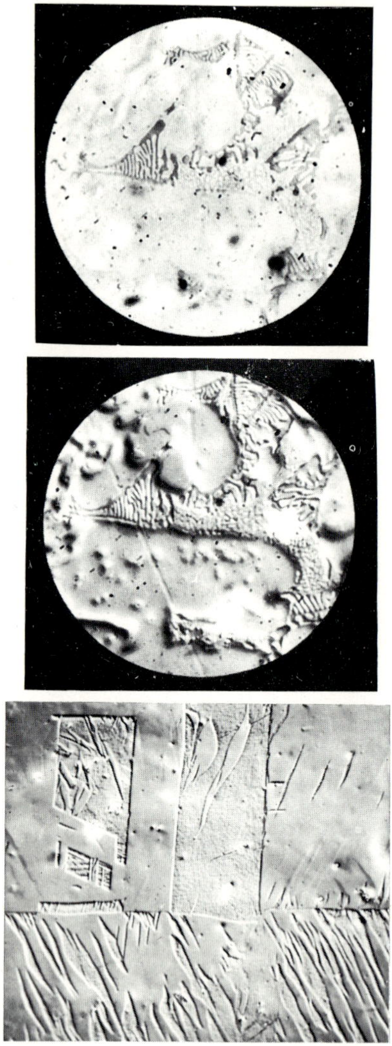

(a)

(b)

(c)

FIG. 30. (a) Cobalt steel viewed with normal reflected light. Magnification ×74. (b) The same area viewed by interference contrast (*Courtesy of Nachet*). (c) Cobalt. Interference contrast. Magnification ×825. (*Courtesy of Reichert*.)

examined with interference contrast, shown in Fig. 30(c), is an example of this effect.

Figure 31(a) shows a multiple-beam interferogram of cleaved mica. The broken contour lines delineate cleavage steps. It should be pointed out that contour fringes are much sharper for multiple-beam than for two-beam interferometry. In fact, in the two-beam case, the dark and light contours are of almost equal thickness. Another use of multiple-beam interferometry is illustrated in Fig. 31(b), which shows the surface contours produced by a diamond pyramid indentor on the surface of a single crystal of tin. The asymmetric lobes are parallel to the *c*-axis and hence indicate anisotropy of plastic deformation. Another interesting feature of this micrograph is the extent of the surface contours which show the area of surface distorted by the indentor.

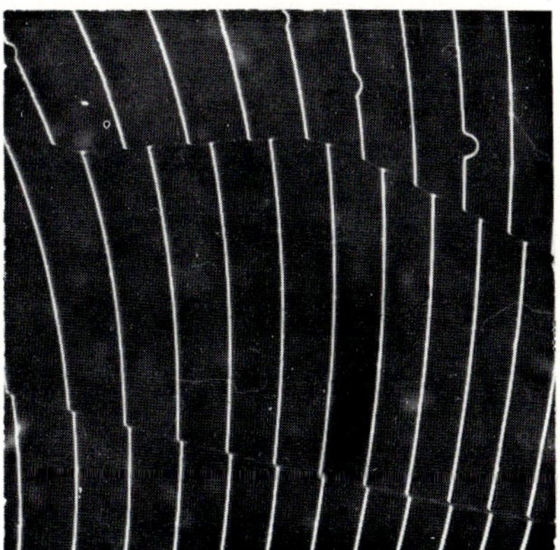

FIG. 31(a). Interferogram of cleaved mica.

An example of polarization interferometry is reproduced in Fig. 32, which is an interferogram of an electro-polished α-brass surface. The displacements of the fringes are a measure of the variations in level to which they correspond.

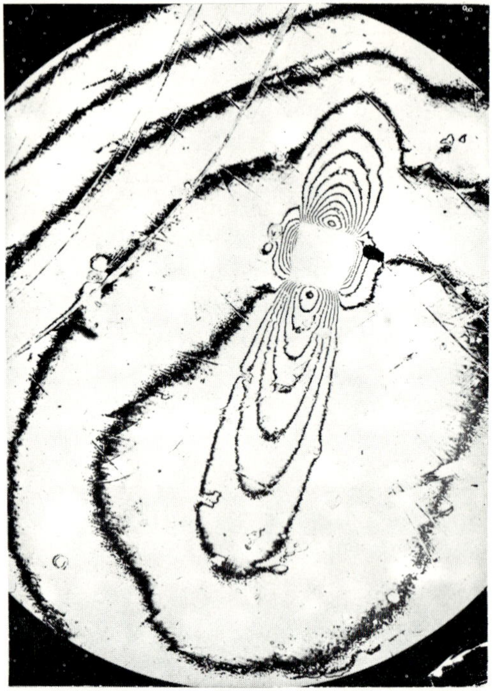

FIG. 31(b). Interference fringes near a Vickers hardness indentation in a tin single crystal. (From S. Tolansky, *International Science and Technology*, Sept. 1962.)

Other metallurgical phenomena which can be studied particularly well using interference techniques include the degree of perfection attained in surface finishing, the breakdown of protective films on corrosion resistant alloys and the nature of slip lines and other surface faults which result from plastic deformation.

On comparing the merits of the techniques for studying surface topography, it is evident that only with interferometric methods can very small step-heights be accurately measured. Unfortunately, interferometric methods are difficult as techniques to use and have the disadvantage of either too low a lateral magnification (×200 for multiple-beam interferometry) or physical instability. In practice, it is often found that a combination of phase contrast and opaque stop is a useful way of obtaining a good idea of the surface topography. The main disadvantage is that neither method permits the actual measurement of step-heights or tilt angles. Probably the best advice to anyone examining surface topography is to use the normal microscope over its full range of magnification (N.B. low magnification means large depth of focus, Chapter 1), and then work through oblique illumination, opaque stop, phase contrast, and finally interferometry if such measurements are required.

FIG. 32. Interferogram taken from an electro-polished surface of α-brass. Magnification ×280. (*Courtesy of Reichert.*)

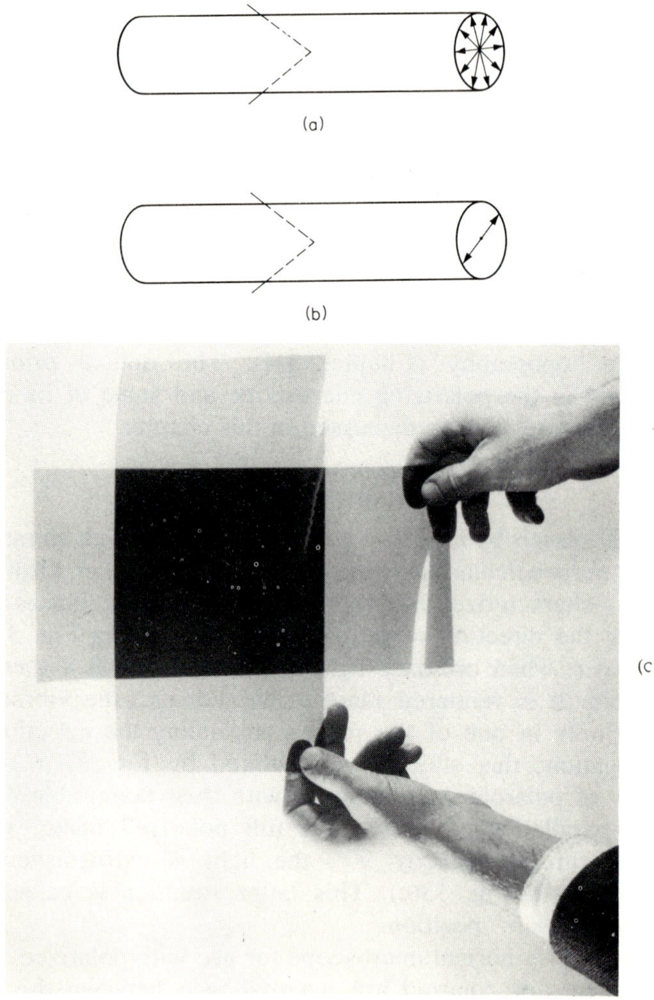

FIG. 33. Vibrations in (a) ordinary and (b) plane polarized light. The arrow indicates the direction of propagation in each case. (c) Extinction observed with two pieces of polaroid in the crossed polars position. (*Courtesy of Cooke, Troughton & Simms.*)

the incident beam is plane polarized when it reaches the specimen, i.e. it is polarized parallel to a particular direction in the polarizer. The analyser, which can be rotated through 360°,

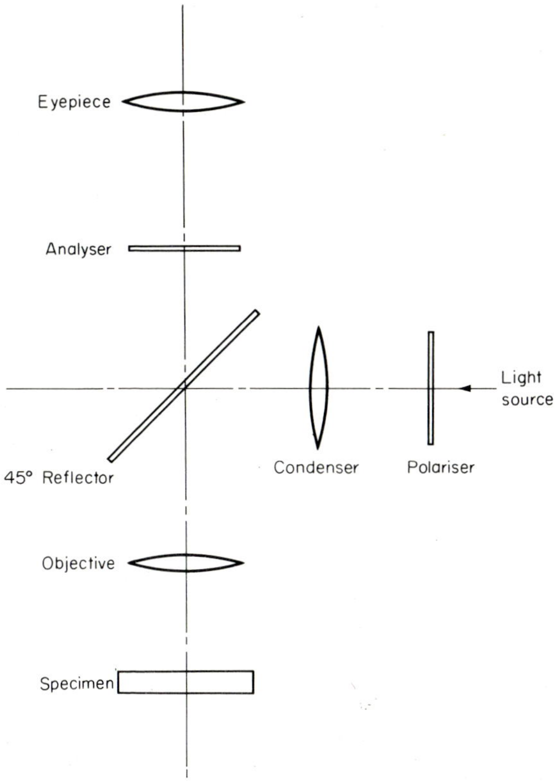

FIG. 34. Polarizing microscope.

is set in a crossed polars relation with respect to the polarizer, i.e. its vibration direction is at right angles to that of the polarizer. Under these conditions, the direct light is extinguished and the field of view appears dark, usually deep violet in colour.

The only other modification to the normal microscope that is necessary is to make the specimen stage capable of a 360°

rotation. It should also be noted, however, that it is necessary for the objective to be completely free from strain and irradiation damage. Strain may result in anomalies from the superimposition of birefringent effects on the state of polarization after reflection by the specimen, and bombardment by particles emitted by radioactive materials disrupts the polarization characteristics of polaroid.

EXAMINATION OF ANISOTROPIC SURFACES

With respect to their behaviour under polarized light, specimen surfaces may be divided into two classes termed isotropic and anisotropic respectively. An isotropic surface reflects a polarized beam without altering the direction of vibration, and consequently appears dark when viewed through crossed polars for all positions of the rotating stage. In the case of an anisotropic surface, the view appears alternately dark and light as the specimen is rotated, there being four positions of complete darkness or extinction in one 360° rotation. The reason for this phenomenon is that the optical properties of an anisotropic surface vary with crystallographic direction. In effect, light reflected from an anisotropic surface is polarized parallel to a certain crystallographic direction in the surface, and hence on reaching the crossed analyser it has been uncrossed by the surface. A position of maximum brightness corresponds to the case where the reflected beam is completely uncrossed, and a position of maximum darkness to that where the reflection produces no uncrossing, i.e. the vibration direction in the incident beam is parallel to that in the surface.

An obvious use of polarized light is in the metallographic examination of polycrystalline anisotropic metals. Since the intensity of polarized light received by the eyepiece depends on the orientation of the crystal reflecting the light, the grain structure will be revealed without etching because the orientation changes from grain to grain. Figure 35(a) and (b) demonstrates this application; when viewed by ordinary light an

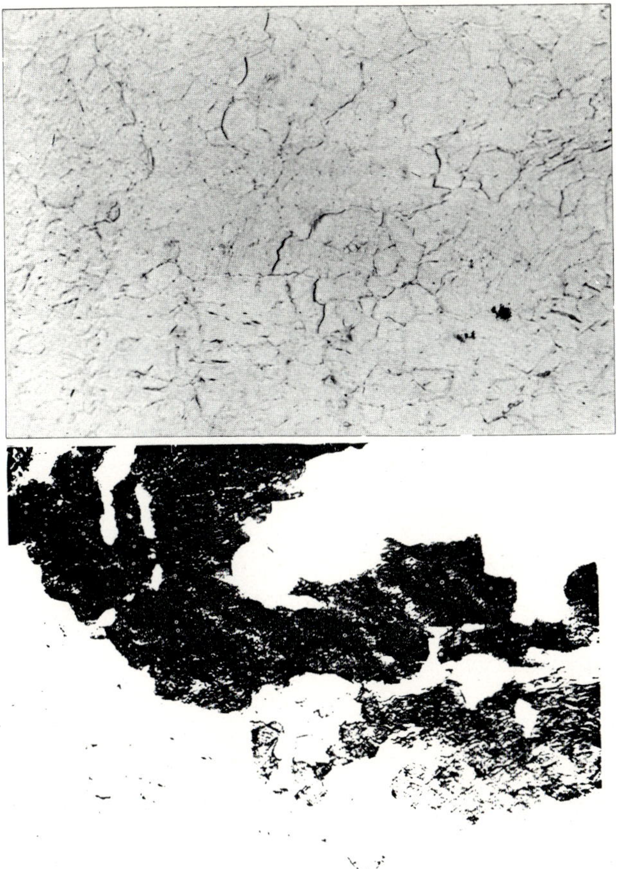

(a)

(b)

FIG. 35. Unetched polycrystalline specimen of tin viewed (a) with ordinary reflected light, and (b) through crossed polars. (*Courtesy of Reichert.*)

unetched specimen of tin appears featureless, but under crossed polars the grain boundaries are revealed because of the variation in polarization direction from grain to grain. This application of polarized light is particularly useful when studying fine details which are likely to be destroyed by etching. An example is given in Fig. 36 and this shows the appearance, under crossed polars, of inclusions in electro-polished α-uranium. An interesting feature observed in the metallography of α-uranium is that it often appears to be optically isotropic when mechanically polished, although its crystal structure is orthorhombic. This observation arises from the effects of plastic deformation in the surface layers. A somewhat similar effect is made use of in transparent crystals to study slip planes (see next section).

FIG. 36. Electro-polished uranium viewed with polarized light. Magnification ×160. (*Courtesy of Reichert.*)

Another application of polarized light is the investigation of preferred orientation in anisotropic polycrystals. Since the setting of the specimen stage for minimum intensity varies with grain orientation, it is possible to estimate the degree of preferred orientation from the fraction of the area of a section having approximately the same extinction angle, i.e., the angle of the specimen stage at which the beam is extinguished. A similar use of the technique is in the study of recrystallization, where differences in orientation between the parent and recrystallized grains of an anisotropic metal show up as differences in intensity.

Other phenomena which can be studied in anisotropic crystals are (1) twinning, which is characterized by sharp changes in orientation, and (2) the distribution of phases in alloys. In multi-phase alloys, it may be difficult to distinguish the phases by etching, especially when they are of similar composition, but since some phases will be isotropic and others anisotropic a distinction is possible with the aid of polarized light.

OTHER USES OF POLARIZED LIGHT MICROSCOPY

The surface of an isotropic metal can be studied with polarized light if it is first covered with an epitaxial film, e.g., of oxide, which is anisotropic. An example of this technique is in the electro-polishing of aluminium, which leads to the formation of an anisotropic anodic oxide film on the surface, and therefore reveals the grain structure of the aluminium substrate.

In addition to its uses in reflection, polarized light finds important applications in transmission. Of particular interest to the physical metallurgist is the study of stress-birefringence in transparent ceramic crystals. An example of this technique is given in Fig. 37, which shows the stress-birefringence pattern seen through a deformed single crystal of rutile, TiO_2, viewed by transmitted light under crossed polaroid. This crystal was deformed in compression and the bright bands delineate slip planes. These bands occur because the strain fields around dislocations alter the orientation of the plane in which the incident beam is polarized, and consequently the light passing through the regions of the crystal adjacent to the slip planes is uncrossed and hence the slip planes appear bright. This technique is therefore extremely useful for studying slip planes in transparent crystals.

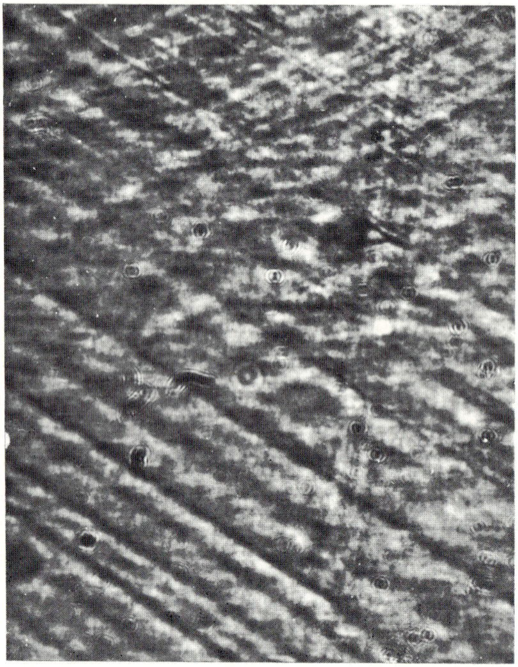

FIG. 37. Stress-birefringence pattern through a single crystal of rutile compressed 3% at 900 C. Magnification ×135. (After Ashbee and Smallman. *Proc. Roy. Soc.* A. **274**, 195 (1963), fig. 2c.)

The use of colour-sensitive filters is another interesting development in polarized light microscopy. These filters have been used to investigate such phenomena as recrystallization with reflected light and plastic deformation with transmitted light in transparent crystals.

SUGGESTIONS FOR FURTHER READING

A. F. HALLIMOND, *Manual of the Polarising Microscope,* Cooke, Troughton & Simms Ltd., 1953.

G. K. T. CONN and F. J. BRADSHAW, *Polarised Light in Metallography,* Butterworths, 1952.

N. H. HARTSHORNE and A. STUART, *Crystals and the Polarising Microscope,* Edward Arnold, 1960.

CHAPTER V

X-RAY METALLOGRAPHY

INTRODUCTION

Diffraction methods are of considerable importance in the structural analysis of metals, since they can give information about the nature of the lattice, i.e., structure-type and lattice parameter, and also details about crystal size, orientation and perfection. In this section we shall outline some of the principles of X-ray diffraction. Electron diffraction will be considered later together with electron microscopy since the combined techniques allow additional structural information to be derived, and neutron diffraction is considered in the chapter dealing with individual atomic behaviour.

WHITE AND CHARACTERISTIC RADIATION

X-rays are a form of electromagnetic radiation having an important difference from light waves in that they have shorter wavelengths ($\lambda \sim 1$ Å). An X-ray beam is emitted when a metal target is bombarded with fast electrons in an evacuated tube, and takes the form of a continuous spectrum of "white" radiation spread over a wide range of wavelengths, together with a superimposed line spectrum characteristic of the metal being bombarded (see Fig. 38). Much of the weak "white" radiation can be filtered out by placing a thin metal foil in the beam (e.g., a nickel filter in a beam of X-rays from a copper target), and the strong characteristic radiation is only slightly attenuated. This filtered radiation is sufficiently monochromatic for most purposes, but pure monochromatic (single wavelength) beams can be produced by impinging the beam on a quartz, lithium fluoride or mica crystal oriented with respect to

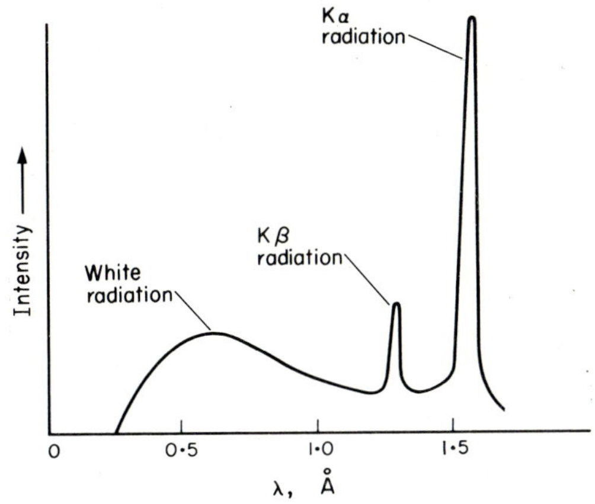

FIG. 38. Schematic diagram showing the distribution of X-ray intensity from a copper target.

the beam so that only the desired wavelength is reflected according to the Bragg law (see below).

THE LAUE CONDITIONS

If a beam of X-rays of wavelength λ impinges on a row of atoms equally spaced a distance a apart, each atom acts as a source of secondary waves, as in ordinary diffraction theory, and these reinforce in certain directions to give rise to zero, first, second and higher order diffracted beams. The condition for reinforcement is that the path-difference for rays scattered by adjacent atoms in the row must be an integral number of wavelengths. If the incident beam makes an angle α_0 and the diffracted beam an angle α with the row of atoms, as shown in Fig. 39, the condition for reinforcement is that

$$a(\cos\alpha - \cos\alpha_0) = h\lambda$$

where h is an integer. This relation is satisfied by cones of

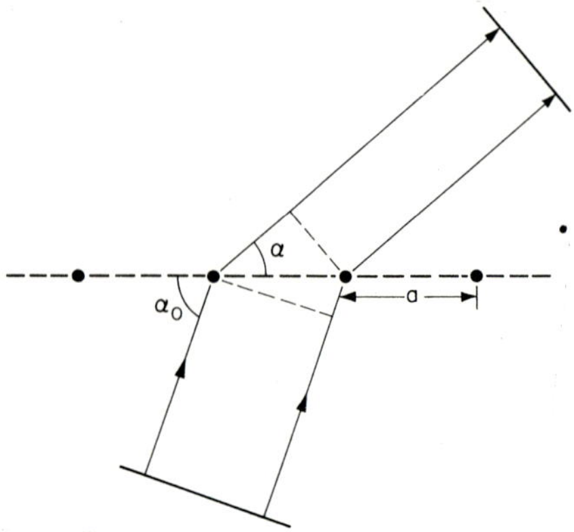

FIG. 39. One-dimensional Laue condition.

diffracted rays of semi-angle α, each cone for one order of diffracted rays.

For a two-dimensional array of atoms with spacing a in the a-direction and b in the b-direction, two conditions have to be satisfied simultaneously for reinforcement of the diffracted rays, namely:

$$a(\cos \alpha - \cos \alpha_0) = h\lambda$$

and

$$b(\cos \beta - \cos \beta_0) = k\lambda$$

where β and β_0 are the corresponding angles of diffraction and incidence for the row of atoms of spacing b, and k is the integer giving the order of reflection for the b-rows. Similarly, for a three-dimensional array of atoms three conditions have to be fulfilled simultaneously for diffraction, viz.

$$a(\cos \alpha - \cos \alpha_0) = h\lambda$$
$$b(\cos \beta - \cos \beta_0) = k\lambda$$
$$c(\cos \gamma - \cos \gamma_0) = l\lambda$$

These equations are known as the Laue equations, and the conditions the Laue conditions. These equations correspond to three sets of cones (i.e., a set around the *a*-axis, a set around the *b*-axis and one around the *c*-axis respectively) and a reinforced diffracted beam will only travel along a common line of intersection of all three sets of cones. Such a common line of intersection will only arise if there are special values for the variables, and hence the Laue conditions impose severe limitations on the number of diffracted beams.

THE BRAGG LAW

The restriction on the diffracted ray discussed above has been demonstrated in a simpler way by Bragg who has shown that the Laue conditions can be expressed in terms of a condition for reflection from the lattice plane (*hkl*). Figure 40 shows a beam of X-rays of wavelength λ, impinging at an angle θ on a set of crystal planes of spacing d. A beam is reflected at the angle θ only if the rays from each successive plane reinforce each other and for this to occur the extra distance that a ray scattered from each successive plane has to travel must be equal to an integral number of wavelengths, $n\lambda$. Thus, in Fig. 40 the ray scattered by the second plane of atoms has to travel further than that scattered by the top plane by the

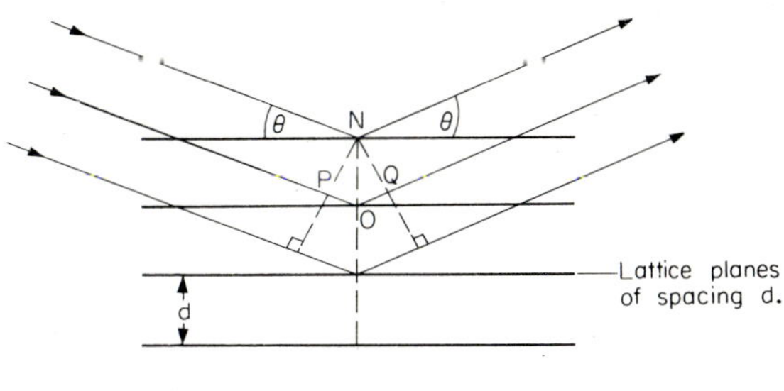

FIG. 40. Conditions for the reflection of *X*-rays from lattice planes.

distance $PO + OQ$ and the condition for reflection and reinforcement will be given by

$$n\lambda = PO + OQ = 2ON \sin\theta = 2d \sin\theta.$$

This is the well-known Bragg law and the critical angular values of θ for which the law is satisfied are known as Bragg angles. It is evident from this equation that the directions of the reflected beams are determined entirely by the geometry of the lattice, and this is governed in turn by the orientation with respect to the beam and spacing of the crystal planes. For a cubic crystal of lattice parameter a, the angles at which the beam is diffracted from the crystal planes *(hkl)* can easily be calculated from the interplanar spacing relationship

$$d_{(hkl)} = a/\sqrt{(h^2 + k^2 + l^2)}.$$

It is conventional to incorporate the order of reflection, n, with the Miller index, and when this is done the Bragg law becomes

$$\lambda = 2a \sin\theta/\sqrt{(n^2h^2 + n^2k^2 + n^2l^2)} = 2a \sin\theta/\sqrt{N}$$

where N is known as the reflection, or line number. Thus the second order reflection from (100) planes is referred to either as the 200 reflection or as line 4. The lattice planes which give rise to a reflection at the smallest Bragg angle are the most widely spaced planes, i.e., $\{100\}$. The next planes in order of decreased spacing will be the $\{110\}$ planes for which $d_{110} = a/\sqrt{2}$, while the octahedral planes will have a spacing equal to $a/\sqrt{3}$. The angle at which any of these planes in a crystal reflect an X-ray beam of wavelength λ may be calculated by inserting the appropriate value of d into the Bragg equation.

The Bragg law is only obeyed for critical values of θ and d. In practice, to ensure that Bragg's law is satisfied and that reflections from various crystal planes will occur, it is necessary to provide a range of either θ or λ values. The various ways in which this can be done leads to the standard methods of X-ray diffraction.

THE STRUCTURE FACTOR

In calculating the intensity scattered by a given atomic structure we have first to consider the intensity scattered by one atom, and then go on to consider the contribution from all the other atoms in the particular arrangement which make up that structure. The efficiency of an atom in scattering X-rays is usually denoted by f, the atomic scattering factor, which is the ratio of A_a, the amplitude scattered by an atom to that scattered by a single electron, A_e. If atoms were merely points, their scattering factors would be equal to the number of electrons they contain, i.e. to their atomic numbers. However, because the size of the atom is comparable with the wavelength of X-rays, scattering from different parts of the atom is not in phase, and the result is that the intensity scattered by an atom $I_a < Z^2 I_e$, where I_e is the intensity scattered by one electron and Z is the atomic number. The atomic scattering factor depends on both the Bragg angle θ and the X-ray wavelength, because the path difference for the individual waves scattered from the various electrons in the atom is zero when $\theta = 0$ and increases with increasing θ; f is therefore expressed as a function of $\sin \theta/\lambda$ as shown in Fig. 41.

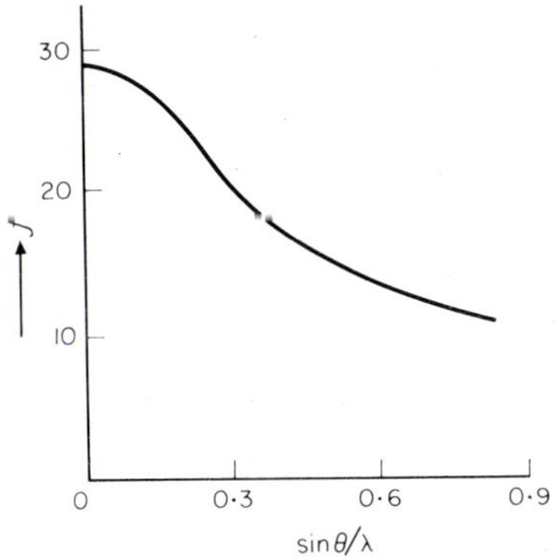

FIG. 41. Variation of atomic scattering factor with $\sin \theta/\lambda$.

To find the intensity scattered by a material of a given structure it is necessary to add waves of different phase and amplitude which arise from atoms of different scattering power situated at different positions in the unit cell. Phase differences are related to the path difference δ between two travelling waves by the relation* $\phi = \delta(2\pi/\lambda)$. Thus, the phase difference between the wave scattered from an atom having co-ordinates x, y, z or fractional co-ordinates u, v and w equal to x/a, y/b, z/c respectively, and that scattered by an atom at the origin, for the *hkl* reflection is given by

$$\phi = 2\pi(hu + kv + lw)$$

The scattered wave is then conveniently expressed in terms of the complex exponential form

$$Ae^{i\phi} = f \exp 2\pi i(hu + kv + lw)$$

since

$$Ae^{i\phi} = A \cos \phi + Ai \sin \phi$$

where A is the amplitude of the wave and $e^{i\phi}$ is the phase factor. The resultant wave scattered by all the atoms of a unit cell is called the structure factor, F, and $|F|$ is the ratio of the amplitude of the wave scattered by all the atoms in the unit cell to the amplitude of the wave scattered by one electron. If a unit cell contains

1, 2, 3 ——— N atoms with fractional co-ordinates u_1, v_1, w_1, u_2, v_2, w_2 ——— and atomic scattering factors f_1, f_2, f_3 ———,

then the structure factor for the *hkl* reflection is given by

$$F_{hkl} = f_1 \exp 2\pi i(hu_1 + kv_1 + lw_1) + f_2 \exp 2\pi i(hu_2 + kv_2 + lw_2)$$
$$+ kv_3 + f_3 \exp 2\pi i(hu_3 + kv_3 + lw_3) + \ldots\ldots$$

*See Appendix to Chapter 5.

or

$$F_{hkl} = \sum_{1}^{N_f} n \exp 2\pi i(hu_n + kv_n + lw_n) \qquad (5.1)$$

the summation extending over all the atoms in the unit cell. Equation (5.1) may be expanded to

$$F_{hkl} = \sum_{1}^{N_f} n \left[\cos 2\pi(hu_n + kv_n + lw_n) + i \sin 2\pi (hu_n + kv_n + lw_n)\right]$$

and the intensity, which is proportional to $|F|^2$ the square of the amplitude of the resultant beam, is obtained by multiplying the expression for F by its complex conjugate,* namely

$$|F|^2 = [f_1 \cos 2\pi(hu_1 + kv_1 + lw_1) + f_2 \cos 2\pi(hu_2 + kv_2 + lw_2) + \_\_\_\_\_\_]^2 + [f_1 \sin 2\pi(hu_1 + kv_1 + lw_1) + f_2 \sin 2\pi(hu_2 + kv_2 + lw_2) + \_\_\_\_\_\_]^2$$

For structures having a centre of symmetry, which includes most metals, the sine terms vanish.

To illustrate the way in which the crystal structure of a metal may be identified from its diffraction pattern,† let us re-examine the intensity scattered from a metal with b.c.c. structure. Since the metal is pure it has identical atoms (i.e., $f_1 = f_2 = f_3$ ______) at the co-ordinates (000) and $(\frac{1}{2}\frac{1}{2}\frac{1}{2})$ and the structure factor equation becomes

*When a wave is expressed in complex form, the square of the amplitude is obtained by multiplying the complex expression for the wave by its complex conjugate, which is obtained by replacing i by $-i$. Thus the complex conjugate of $Ae^{i\phi}$ is $Ae^{-i\phi}$ and $|Ae^{i\phi}|^2 = Ae^{i\phi} Ae^{-i\phi} = A^2$ or $A(\cos \phi + i \sin \phi)\, A(\cos \phi - i \sin \phi) = A^2(\cos^2 \phi + \sin^2 \phi) = A^2$.

†Anomalies arise in electron diffraction due to the occurrence of forbidden reflections. Consequently, X-ray diffraction is usually preferable for establishing crystal structures.

$$I \propto |F|^2 = f^2[\cos 2\pi(0) + \cos 2\pi(h/2 + k/2 + l/2)]^2$$
$$= f^2[1 + \cos \pi(h+k+l)]^2$$

It therefore follows that the intensity I is zero for reflections for which $(h + k + l)$ is an odd number, i.e., these reflections are absent. A physical interpretation of these systematic absences for a metal with b.c.c. structure is as follows. The first reflection which should occur is the first order reflection from the (001) planes, as shown schematically in Fig. 42(a). A first order reflection implies that there is a phase-difference

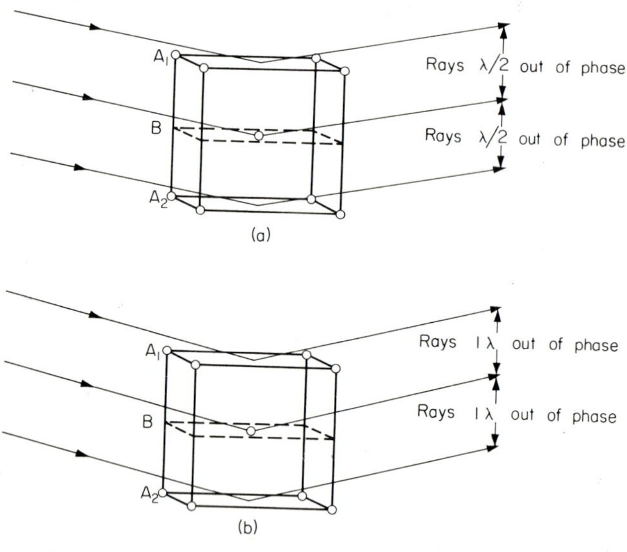

FIG. 42. The origin of absent reflections: (a) showing the cancellation of the 100 reflection and (b) the reinforcement of the 200 reflection from a b.c.c. lattice.

equal to one wavelength between the rays reflected from successive (001) planes, A_1 and A_2. However, rays will also be reflected from the layer of atoms situated half way between successive (001) planes, labelled B, and since these are $\lambda/2$ out of phase with rays from the A-planes they will interfere destructively with the latter. Consequently, for a b.c.c. structure line 1 is absent.* The second-order reflections from the (001) planes, i.e., the 002 reflection, will be present, of course, because in this case the ray from the B-planes will have a phase-difference of exactly one wavelength with that scattered from the A-planes. Complete reinforcement occurs (see Fig. 42b), and line 4 is present. Similar reasoning shows that the first-order reflection, from the (111) planes will be extinguished, while the second-order reflection, i.e., $N = 12$, is not. In the same way, it can be shown that the complete X-ray diffraction pattern of a b.c.c. metal contains the following reflections: $N = 2,4,6,8,10,12$...... or in terms of Miller indices 110,200,211,220,310,222 etc. In general terms, reflections are absent for which $(h + k + l)$ is an odd number.

The X-ray diffraction pattern for an f.c.c. metal can be shown to contain only the lines $N = 3,4,8,11,12,16,19,$ 20.... Thus reflections are absent for which $(h + k),(k + l),$ $(l + h)$ are odd. It is evident that an examination of the absent reflections on an X-ray diffraction pattern reveals the crystal structure of the metal studied.

THE LAUE METHOD AND ORIENTATION DETERMINATION

In the single crystal Laue method the specimen is stationary and is exposed to a beam of white radiation. The variable necessary to ensure that the Bragg law is satisfied for all planes is provided by the range of wavelengths in the beam, i.e., each set of crystal planes chooses the appropriate λ from

*In an ordered alloy, alternate planes A and B are not statistically identical and while the diffracted rays from the A-planes are completely out of phase with those from the B-planes, their intensities are not equal, and hence a weak reflection from line 1, for example, results. These weak reflections, which for a b.c.c. metal occur when $(h + k + l)$ is an odd number, are known as superlattice lines.

the white spectrum to give a Bragg reflection. In the experimental arrangement shown in Fig. 43 either a transmission photograph or back-reflection photograph may be taken, and the diffraction pattern gives rise to spots which lie on ellipses in the transmission case or hyperbolae in the back-reflection case. All the spots on any ellipse or hyperbola represent

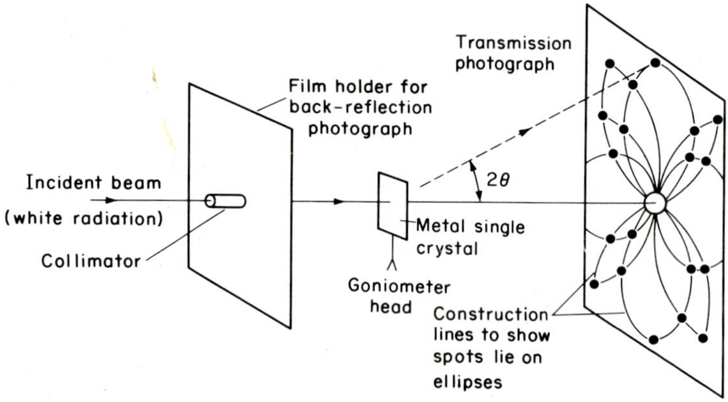

FIG. 43. Schematic diagram of Laue diffraction method. (From R. E. Smallman, *Modern Physical Metallurgy,* fig. 21a.)

reflections from planes of the same zone. For this reason a Laue pattern indicates the symmetry of the crystal. Thus, if the beam is directed along a <111> or <100> direction of a cubic crystal, the Laue pattern will show three-fold or four-fold symmetry respectively. The most important metallurgical use of the Laue method is in the determination of the orientation of single crystals. The method, explained in detail below, consists essentially of plotting spots taken from the film onto a stereogram, and comparing the angles between them with the known interplanar angles for the crystal structure. Figure 44(a) represents an actual Laue back-reflection photograph of a cubic crystal* from which it is evident that the spots, each representing a different family of planes, lie on well-defined hyperbolae. To index the spots and hence determine the orientation of the crystal, only the strongest spots should be

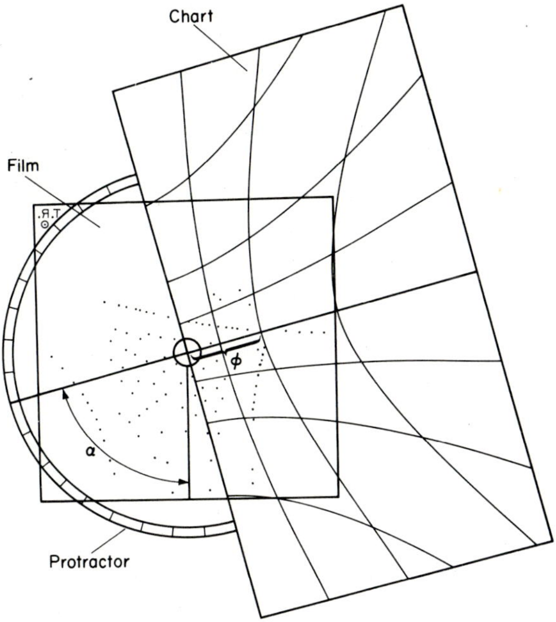

FIG. 44(a). The use of the Greninger chart in the determination of crystal orientation

considered at first, since these are formed by reflection from prominent low index planes. Such spots usually lie at the intersections of many hyperbolae. It is evident in Fig. 43 that the distance between a spot and the centre of the film is a

*N.B. An identification mark must be made on the film which bears reference to the geometrical optics of the apparatus. In Fig. 44(a) the exposed film has been marked before developing with the symbol T.R., the letters denoting top right and the arrow head symbol signifying that the X-rays from the source came through the central hole towards the observer as he looks at the film.

measure of the angle between the crystal plane to which the spot corresponds and the photographic film. Greninger has plotted a chart from which this angle can be read directly and simply by superimposing the chart on the film as illustrated in Fig. 44(a). Exact reproductions of Greninger's original chart, suitable for a specimen-to-film distance of 3 cm, are commercially available.* Each spot to be indexed is marked in ink on the film by a reference letter, a,b,c,..., and transferred to a stereographic projection. In carrying out this operation, the film is regarded as the plane of the stereogram and consequently a spot in the top right-hand corner of the film has to be transferred to the top right-hand segment of the stereogram, etc. To make such a transfer, it is necessary to measure the "azimuthal" angle, α, between a reference line drawn in ink on the film from, say, north to south through the central hole and the line subtended at the central hole by the spot. This angle, α, is measured with a protractor; in practice it is convenient to fix a protractor to the bottom of the Greninger chart as shown in Fig. 44(a). On the stereogram, shown in Fig. 44(b), a line is drawn between the centre and south pole, and from it is measured the azimuthal angle, α. The Greninger chart is superimposed on the film and rotated in such a way to give the azimuthal angle, as shown in Fig. 44(a), and then the angle ϕ between the film and the crystal plane corresponding to the spot concerned is read off along the vertical axis of the chart. This angle is then transferred to the stereogram. This is carried out with the stereogram rotated through the azimuthal angle over the Wulff net, and the required angle ϕ is measured from the centre along the axis between the north and south poles.

After plotting, say, 20 strong spots on the stereogram, the Wulff net is used to measure the angles between them. Then, using a table of interplanar angles tentative indices are assigned to these spots. Eventually, a self-consistent set of indices should be obtained. The index of the centre of the stereogram

*Address: Institute of Physics and the Physical Society, 47 Belgrave Square, London, S.W.1.

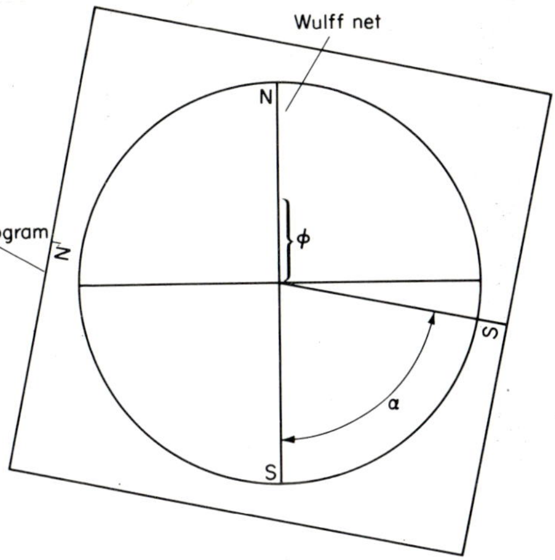

FIG. 44(b). The use of the Greninger chart in the determination of crystal orientation.

is then the index of the pole of the crystal plane perpendicular to the X-ray beam, and the north pole of the stereogram is the index of the horizontal crystal plane.

It is now possible to rotate the crystal, by adjusting the two arcs of the goniometer, to bring any plane into any desired position relative to the X-ray beam or alternatively to a reference line marked on the base of the goniometer. The two angles through which the crystal must be rotated to bring a given pole to the centre of the stereogram can be measured by the Wulff net from the north to south line and equator of the stereogram. To make the required adjustment to the goniometer arcs, it is necessary to look along the direction of the X-ray beam and, regarding the crystal as a mirror, bring the desired spot into the centre of the film by rotating through the same angles previously measured on the stereogram. The axes of rotation are now vertical and horizontal.

All the spots lying on a given hyperbola of a back-reflection photograph represent planes belonging to the same zone, and an alternative way of determining the orientation is to plot on the stereogram the axes of zones rather than the poles of individual planes. The Greninger chart can then be rotated over the film until one of its horizontal hyperbolae coincides with the row of spots being measured, and ϕ, the angle from the centre of the chart to the row of spots, is the complement of the angle between the zone axis and the X-ray beam. Thus, to transfer the zone-axis to the stereogram, ϕ is measured from the circumference of the Wulff net. As before, the stereogram first has to be rotated over the net through the azimuthal angle α, and ϕ is measured along the north to south great circle.

By convention, metallurgists usually work with a stereographic projection of the poles of planes and not of zone axes. Consequently, it is usual to determine orientations from back-reflection films by the former method and not by the latter, although both methods are in general use.

In recent years the use of the Laue technique has been extended to the study of imperfections resulting from crystal growth or deformation, because it is found that the Laue spots from perfect crystals are sharp, while those from imperfect or deformed crystals are elongated. This elongated appearance of the diffraction spots is known as asterism and it arises in a manner analogous to the formation of caustic curves by the reflection of light from curved mirrors.

POWDER METHOD AND THE ACCURATE MEASUREMENT OF LATTICE PARAMETERS

The powder method, devised independently by Debye and Scherrer, is probably the most generally useful of all the X-ray techniques. Monochromatic radiation is used, and the range of θ values required to satisfy Bragg's law is ensured by having a specimen in the form of either an extrusion of bonded filings or a polycrystalline wire with very small grain size. Such a collection of randomly oriented crystals will contain sufficient particles with the correct orientation to permit Bragg reflections from each of the possible reflecting planes for all angles of rotation, i.e., the specimen is rotated about its cylindrical axis and a powder pattern results from, as it were, a series of superimposed rotating crystal patterns. The angle between the direct X-ray beam and the reflected ray is 2θ, and consequently each set of crystal planes gives rise to a cone of reflected rays of semi-angle 2θ, where θ is the Bragg angle for that particular set of reflecting planes producing the cone. Thus, if a film is placed around the specimen, as shown in Fig. 45(a), the successive diffracted cones intersect the film

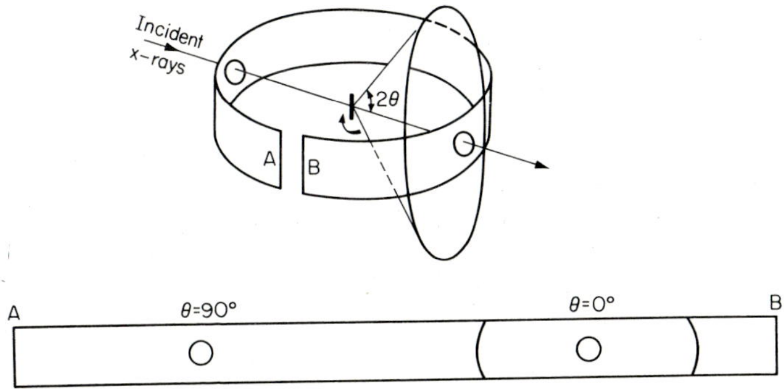

FIG. 45(a). Schematic diagram of the powder method. (From R. E. Smallman, *Modern Physical Metallurgy,* fig. 22b.)

to produce curves concentric with the entrance and exit holes. Fig. 45(b) shows a powder photograph of molybdenum, a typical b.c.c. metal, from which it can be seen that the lines are fairly evenly spaced.

Applications of the powder method depend on the accurate measurement of either line position or line intensity. The arrangement of the lines in any diffraction pattern is characteristic of the material being examined, and hence an important practical use of the method is in the identification of unknown

phases. Thus, the Bragg law dictates the positions of the reflected beams since these are determined by the crystal structure and lattice parameter(s) of each phase. The intensities of the reflected beams are determined not by Bragg's law, i.e., not by the type and size of the unit cell, but by the distribution of atoms within it (see the section under Structure Factor). Consequently, accurate measurement of diffracted intensities may be used to determine these atomic positions.

Probably the most common use of the powder method is in the accurate determination of lattice parameters. From the Bragg law we have the relation $a = \lambda\sqrt{N}/2 \sin\theta$ which, because both λ and N are known and θ can be measured for the appropriate reflection, can be used to determine the lattice parameter of a material. Several errors are inherent in the method, however, and the most common include shrinkage of the film during processing, eccentricity of the specimen in the camera, and absorption of the X-rays in the sample. These errors affect the high-angle diffraction lines least and, consequently, the most accurate parameter value is given by determining a value of a from each diffraction line, plotting it on a graph against an angular function of the $\cos^2\theta$-type and then extrapolating the curve to $\theta = 90°$. The determination of precise lattice parameters is of importance in many fields of physical metallurgy, particularly in the study of phase boundaries, thermal expansion coefficients, density determination, the variation of properties with alloy composition, precipitation from solid solution, and the evaluation of thermal stresses.

FIG. 45(b). Powder photograph of molybdenum taken with copper $K\alpha$ radiation in a Phillips camera (5·73 cm radius).

In addition to photographic recording, the diffracted X-ray beam may be detected directly using a counter tube (either Geiger, proportional or scintillation type) with associated electrical circuitry. A schematic arrangement of an X-ray diffractometer is shown in Fig. 46(a). A divergent beam of

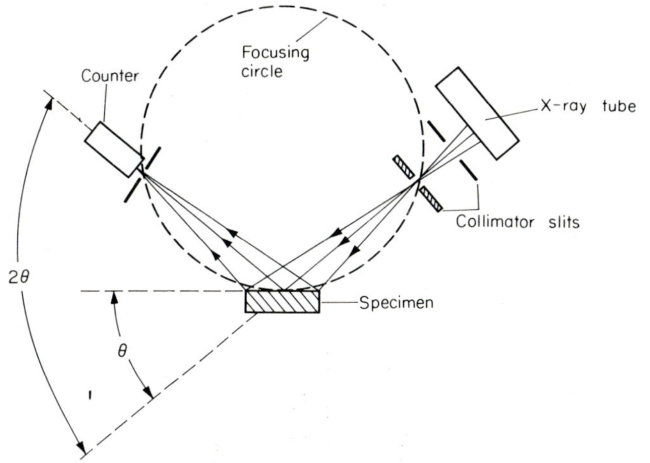

FIG. 46(a). Geometry of a conventional diffractometer.

filtered, or monochromatized, radiation impinges on the flat face of a powder specimen. This specimen is rotated at precisely one-half of the angular speed of the receiving slit so that a constant angle between the incident and reflected beams is maintained. The receiving slit is mounted in front of the counter on the counter tube arm, and behind it is usually fixed a scatter slit to ensure that the counter receives radiation only from that portion of the specimen exposed to the primary beam. The intensity diffracted at the various angles is automatically recorded on a chart of the form shown in Fig. 46(b) and this can be quickly analysed for the appropriate θ and d values.

An important use of precision lattice parameters is in the

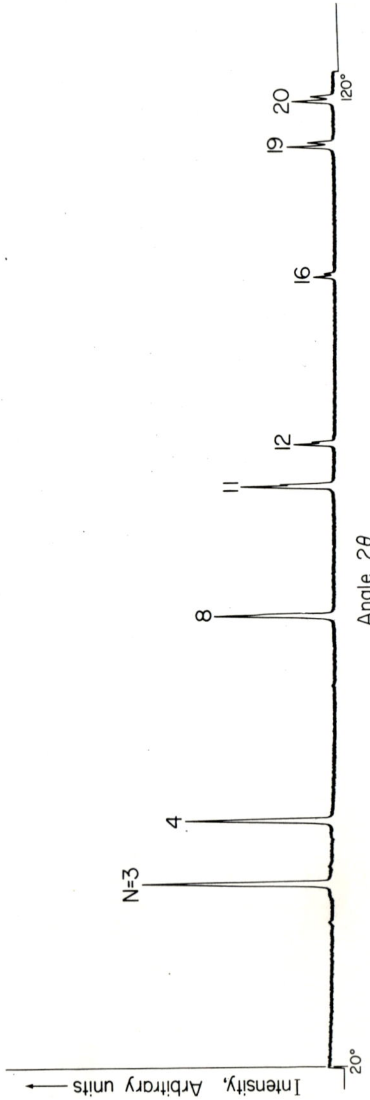

FIG. 46(b). Geiger counter trace of powdered aluminium.

identification of inorganic substances using a card index file compiled by the ASTM. All the substances ever studied by X-ray powder diffraction are filed in order of their largest *d*-spacing. Consequently, a substance can be identified by measuring its *d*-spacings and checking these against ASTM cards having approximately the same largest *d*-spacings.

Appendix to Chapter V

THE RECIPROCAL LATTICE AND THE REFLECTION SPHERE

It is evident from the discussion of the Bragg law given already that to determine the conditions for diffraction, consideration must be given to the geometry of sets of crystal planes. Such a consideration often becomes quite involved and to simplify the more complex diffraction problems, use is made of the reciprocal lattice concept in which the sets of lattice planes are represented more simply by a set of points.

The reciprocal lattice is constructed from the real lattice by drawing a line from the origin normal to the lattice plane *hkl* under consideration of length, d^*, equal to the reciprocal of the interplanar spacing d_{hkl}. The construction of part of the reciprocal lattice from a face-centred cubic crystal lattice is shown in Fig. 47. It can be seen that included in the reciprocal lattice are the points which correspond not only to the true lattice planes with Miller indices (*hkl*) but also to the fictitious planes (*nh*, *nk*, *nl*) which give possible X-ray reflections. In a sense, the reciprocal lattice corresponds to the spectrum possible from the crystal lattice. Thus, because a particular lattice type is characterized by "absent" reflections the corresponding spots in the reciprocal lattice will also be missing, and it can be deduced that an f.c.c. Bravais lattice is equivalent to a b.c.c. reciprocal lattice, and vice-versa.

A simple geometrical construction in the reciprocal lattice gives the conditions that correspond to Bragg reflection. Thus, if a beam of wavelength λ is incident on the origin of the reciprocal lattice, then a sphere of radius $1/\lambda$ drawn through the origin will intersect those points which correspond to the

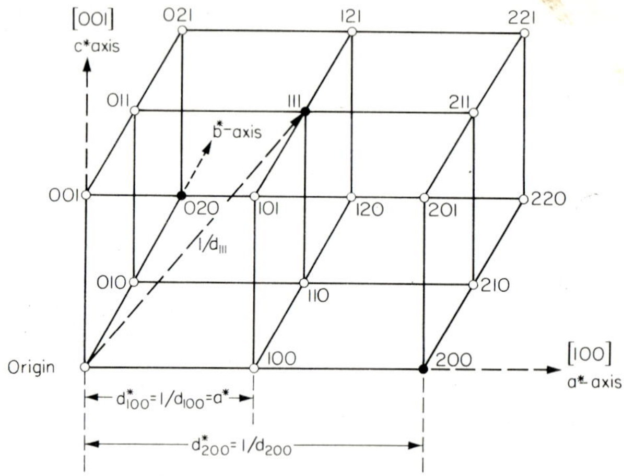

FIG. 47. The reciprocal lattice of a f.c.c. structure. The closed circles represent the reciprocal lattice points which give rise to diffraction.

reflecting planes of a stationary crystal. This can be seen from Fig. 48 in which the reflecting planes *AB* have a reciprocal point at d^*. If d^* lies on the surface of the sphere of radius $1/\lambda$ then

$$d^* = \frac{1}{d_{hkl}} = 2 \sin \theta/\lambda$$

and the Bragg law is satisfied. It will be evident that at any one setting of the crystal, few, if any, points will touch the sphere of reflection. This is the condition mentioned in the previous section, for a stationary single crystal and a monochromatic beam of X-rays, when the Bragg law is not obeyed except by chance. To ensure that the Bragg law is satisfied the crystal has to be rotated in the beam and, since this corresponds to a rotation of the reciprocal lattice about the origin, each point must pass through the reflection surface. The corresponding reflecting plane reflects twice per revolution.

To illustrate this feature let us re-examine the powder method. In the powder specimen the number of crystals is sufficiently large that all possible orientations are present and, in terms of the reciprocal lattice construction, we may suppose that the reciprocal lattice is rotated about the origin in all

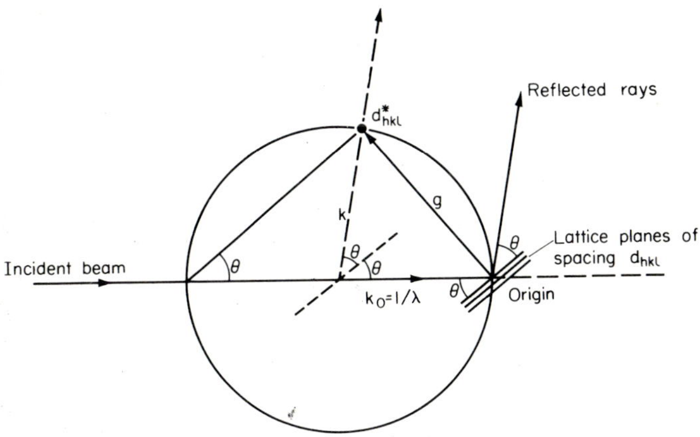

FIG. 48. Ewald sphere construction for diffraction maxima.

possible directions. The locus of any one lattice point during such a rotation is of course a sphere. This locus-sphere will intersect the sphere of reflection in a small circle about the axis of the incident beam as shown in Fig. 49, and any line joining the centre of the reflection sphere to a point on the small circle is a possible direction for a diffraction maximum. This small circle corresponds to the powder halo discussed previously. From this figure it can be seen that the radius of the sphere describing the locus of the reciprocal lattice point (hkl) is $1/d_{(hkl)}$ and that the angle of deviation of the diffracted

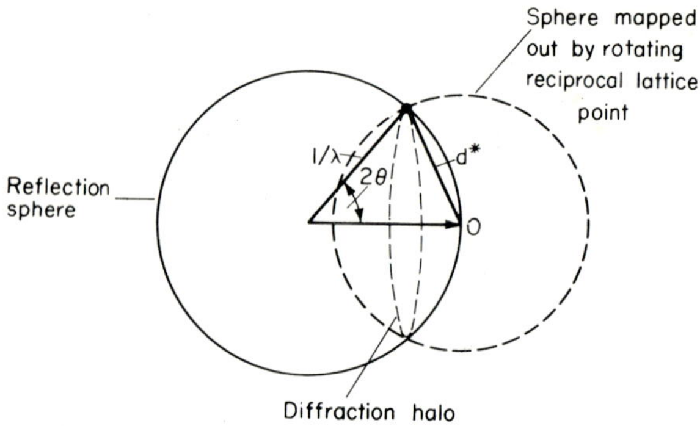

FIG. 49. Reciprocal lattice construction for powder diffraction haloes.

beam 2 θ is given by the relation

$$(2/\lambda)\sin\theta = 1/d_{(hkl)}$$

which, of course, is the Bragg condition.

VECTOR NOTATION

It is usual to describe lengths and directions in the reciprocal lattice in terms of a vector notation. If **a**, **b**, and **c** are the edge dimensions of the real structure cell, then the lengths of the edges of the reciprocal lattice cell are respectively **a***, **b*** and **c***. The reciprocal lattice axis **a*** is, of course, perpendicular to the axes **b** and **c** (i.e., perpendicular to the (100) plane which contains **b** and **c**), the axis **b*** is perpendicular to **c** and **a**, and **c*** to **a** and **b**. These relations are summed up in the vector equations

$$\mathbf{a}\,.\,\mathbf{b} = \mathbf{a}\,.\,\mathbf{c} = \mathbf{b}\,.\,\mathbf{a} = \mathbf{b}\,.\,\mathbf{c} = \mathbf{c}\,.\,\mathbf{a} = \mathbf{c}\,.\,\mathbf{b} = 0$$

and

$$\mathbf{a}\,.\,\mathbf{a}^* = \mathbf{b}\,.\,\mathbf{b}^* = \mathbf{c}\,.\,\mathbf{c}^* = 1$$

where the dot product denotes the scalar product of two vectors, and is equal to the product of the magnitude of the two vectors and the cosine of the angle between them (e.g., $\mathbf{a.b} = ab \cos\theta = ab \cos 90° = 0$, and $\mathbf{a.a^*} = a\frac{1}{a}\cos 0° = 1$).

A lattice point in the real cell may be defined by the vector

$$\mathbf{r} = u\mathbf{a} + v\mathbf{b} + w\mathbf{c}$$

and a point in the reciprocal lattice by

$$\mathbf{r}^*(hkl) = h\mathbf{a}^* + k\mathbf{b}^* + l\mathbf{c}^*$$

which must be normal to the primary lattice planes with Miller indices (hkl) and of magnitude $1/d_{hkl}$. When the reciprocal lattice point $\mathbf{r}^*$ coincides with that for which diffraction occurs this fact is indicated by designating the reciprocal lattice vector with the symbol $\mathbf{g}$.

DIFFRACTION

Diffraction from a crystal can be discussed in terms of the reciprocal lattice. In considering diffraction it is necessary to calculate the phase difference between the rays scattered from any two lattice points such as A_1 and A_2 in Fig. 50(a). The distance between the points is defined by the vector $\mathbf{r}$, the incident beam by the vector $\mathbf{k}_0$, and the diffracted beam by the vector $\mathbf{k}$. The path difference between the two scattered waves is

$$\begin{aligned} d &= \mathbf{r}\,.\,\mathbf{k} - \mathbf{r}\,.\,\mathbf{k}_0 \\ &= \mathbf{r}\,.\,(\mathbf{k} - \mathbf{k}_0) \\ &= \mathbf{r}\,.\,\mathbf{K} \end{aligned} \qquad (1)$$

The vector $\mathbf{K} = (\mathbf{k} - \mathbf{k}_0)$ has the following simple geometrical meaning. Referring to the vector diagram shown in Fig. 50(b), it is evident that $\mathbf{K}$ defines the normal of the plane which reflects the direction of incidence $\mathbf{k}_0$ into the direction of diffraction $\mathbf{k}$.

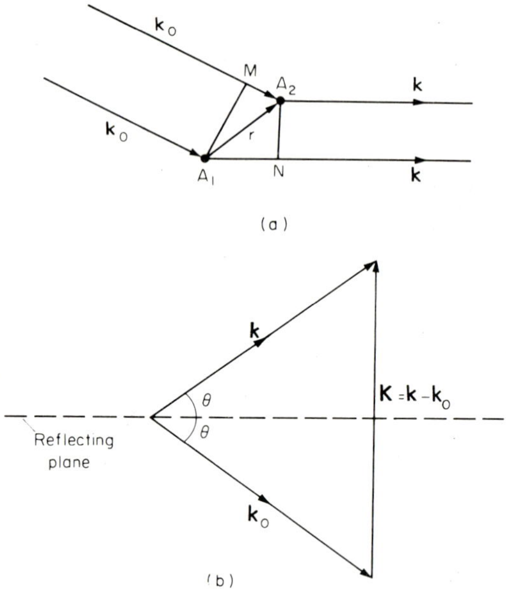

FIG. 50. Vector diagram showing relation between $\boldsymbol{k}_0$ and $\boldsymbol{k}$.

Defining the glancing angle of incidence (= glancing angle of reflection) as θ, we see that since $\mathbf{k}_0$ and $\mathbf{k}$ are of unit magnitude,

$$\mathbf{K} = 2 \sin \theta$$

The phase difference ϕ is the path difference multiplied by $2\pi/\lambda$, where λ is the X-ray wavelength, and therefore

$$\phi = (\mathbf{r} \,.\, \mathbf{K})\frac{2\pi}{\lambda} \tag{2}$$

The vector $\mathbf{K}/\lambda$ can be expressed in terms of the reciprocal lattice as

$$\frac{\mathbf{K}}{\lambda} = \frac{\mathbf{k} - \mathbf{k}_0}{\lambda} = h\mathbf{a}^* + k\mathbf{b}^* + l\mathbf{c}^* \tag{3}$$

and therefore the phase difference becomes

$$\phi = 2\pi(h\mathbf{a}^* + k\mathbf{b}^* + l\mathbf{c}^*).(u\mathbf{a} + v\mathbf{b} + w\mathbf{c})$$
$$\phi = 2\pi(hu + kv + lw) \qquad (4)$$

The Laue equations can be derived from equation (3). For a periodic distribution of scattering points along the a-axis the Laue condition is obtained by taking the scalar product of each side of the equation with the vector **a**, so that

$$\frac{\mathbf{a}\,.\,\mathbf{K}}{\lambda} = \mathbf{a}\,.\,(h\mathbf{a}^* + k\mathbf{b}^* + l\mathbf{c}^*) = h$$

The three Laue equations are respectively

$$\mathbf{K}\,.\,\mathbf{a} = 2a \sin\theta \cos\alpha = h\lambda$$
$$\mathbf{K}\,.\,\mathbf{b} = 2b \sin\theta \cos\beta = k\lambda$$
$$\mathbf{K}\,.\,\mathbf{c} = 2c \sin\theta \cos\gamma = l\lambda$$

and these have the following simple geometrical interpretation. Referring to Fig. 51(a), it can be seen that cos α, cos β and

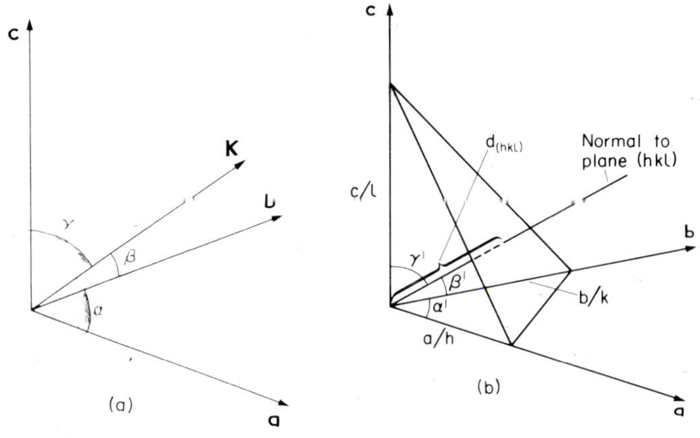

FIG. 51. Diagram showing the relation between the vector **k** and the lattice vectors.

$\cos\gamma$ are the direction cosines of **K**, the normal to the reflecting plane. According to the above equations, $\cos\alpha \propto h/a$, $\cos\beta \propto k/b$ and $\cos\gamma \propto l/c$ for the plane (*hkl*) shown in Fig. 51(b). The direction cosines of the normal to this plane are $\cos\alpha' \propto h/a$, $\cos\beta' \propto k/b$ and $\cos\gamma' \propto l/c$. These are identical to the direction cosines of the normal to the reflecting plane, i.e., $\alpha = \alpha'$, $\beta = \beta'$ and $\gamma = \gamma'$. Consequently the plane (*hkl*) and the reflecting plane are one and the same. This means that when the Laue conditions are satisfied the scattered intensity is a maximum, and the incident beam is related to the scattered beam such that one may be considered as derived from the other by reflection in the plane (*hkl*). The Laue equations imply that

$$\frac{a}{h}\cos\alpha = \frac{b}{k}\cos\beta = \frac{c}{l}\cos\gamma = \frac{\lambda}{2\sin\theta}$$

and, since in Fig. 51(b), $\cos\alpha = d_{(hkl)} \cdot h/a$ etc., i.e., $a/h \cos\alpha = d_{(hkl)}$, then

$$d_{(hkl)} = \frac{\lambda}{2\sin\theta}$$

and therefore $$2d_{(hkl)}\sin\theta = \lambda$$

which is Bragg's law.

SUGGESTIONS FOR FURTHER READING

R. W. James, *The Optical Principles of the Diffraction of X-rays.*

C. S. Barrett, *The Structures of Metals.*

A. Taylor, *An Introduction to X-ray Metallography,* Chapman & Hall, 1952.

CHAPTER VI

SPECIALIZED X-RAY DIFFRACTION TECHNIQUES

INTRODUCTION

In addition to orientation determination by Laue back reflection and precision lattice parameter measurement by the powder method, there are several more specialized X-ray techniques which are used in some metallurgical laboratories; these include preferred orientation determination, small-angle scattering and micro-probe analysis. X-ray diffraction techniques have also been developed to study the nature of dislocations in imperfect crystals.

PREFERRED ORIENTATION

Three sources of preferred orientation or texture are of importance to metallurgists. These are summarized as follows:

1. Plastic deformation processes such as rolling and extrusion, in which individual grains tend to rotate into a common orientation relationship with the rolling direction or wire axis. Such deformation textures develop gradually with strain and become quite marked above about 60% strain.
2. Recrystallization which usually produces a texture different from that of the parent deformation texture.
3. The oriented growth of surface films, e.g., during corrosion or electro-polishing.

The degree of preferred orientation is easily detected on a monochromatic X-ray transmission picture; because the grains are not randomly oriented, and hence the Debye ring is not uniform in intensity but is broken up into intense arcs where the grains of similar orientation reflect (see Fig. 53).

WIRE TEXTURES

In cold drawn or swaged rods all the grains are usually aligned with a specific crystallographic direction [UVW] approximately parallel to the wire axis. Such a preferred orientation is referred to as a fibre or wire texture. When a fibre-textured wire is examined with X-rays the limited number of grain orientations that are present in the wire give rise to "spotty" maxima on the Debye rings instead of a uniform intensity as arises when the metal has a random orientation. While the grains have a common [UVW] axis, all other axes have a random distribution around the wire and hence all azimuthal orientations around the axis are equally probable.

A common texture found in metals is for the [111] direction in each grain to be oriented parallel to the wire axis. If an X-ray transmission photograph is taken of such a wire it is a simple process to draw a stereographic projection of the poles of the reflecting planes. The relation between the (hkl) reflecting planes of the textured wire, the diffracted X-ray spots and their stereographic projection is shown in Fig. 52(a). For the (100) reflecting planes, for example, the stereographic projection or pole-figure, is shown schematically in Fig. 52(b). All reflections fall on the projection at equal angles α from the projection of the fibre axis and, because of the Bragg condition, on a reflection circle which lies at $90° - \theta$ from the position of the incident beam. Since the indices of the (hkl) plane giving rise to the spots are known, it is a simple matter to deduce the indices of the fibre axis by measuring the angle between a spot and the fibre axis and comparing this value with the computed angles between crystallographic directions.

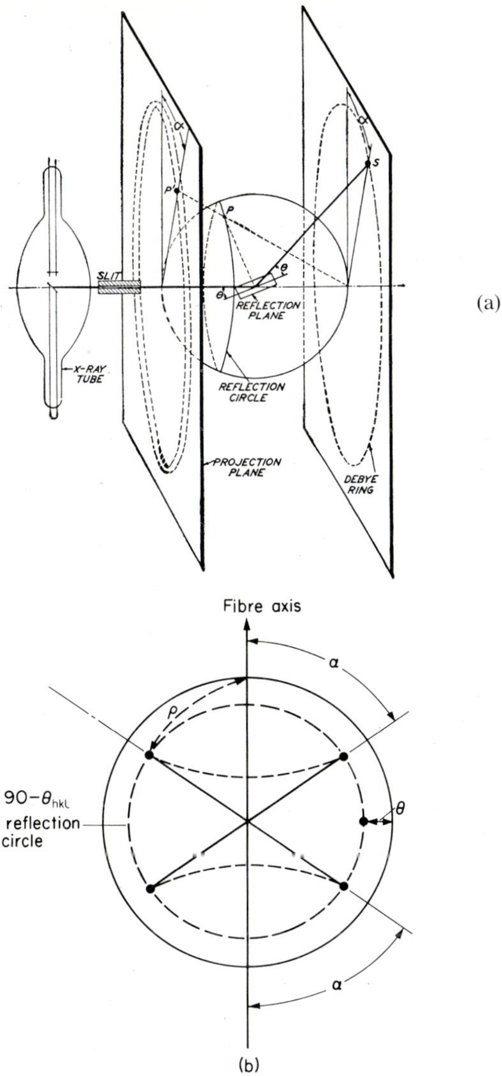

FIG. 52. Diagram showing the relation between (a) the stereographic projection and the diffraction pattern and (b) fibre axis and the stereographic projection. (After C. S. Barrett, *Structure of Metals,* p. 171. fig. 2, and p. 172, fig 3.)

SHEET TEXTURES

Although the orientation of all the grains will be close to some ideal orientation there is a spread in orientations which is best represented by a stereographic projection. For rolling textures, the centre of the projection usually corresponds to the sheet normal, the North-South direction to the rolling direction and the East-West direction to the transverse direction of the sheet. The distribution of a particular set of planes, such as the {111} planes in a f.c.c. metal, is then indicated by contours representing the density of poles of this set of planes at various points on the diagram, or pole-figure as it is called. The peaks on the pole-figure are often represented by ideal orientations, i.e., the peaks correspond to the positions at which the poles of particular set of (*hkl*) planes would occur if the sheet were a single crystal in the ideal orientation. Thus the (111) pole-figure of Fig. 56 (a) may be described by the ideal orientation {110} ⟨112⟩, which means that the position of {111} peaks on the pole-figure can be accounted for if a plane of the form {110} lies parallel to the rolling plane and a direction of the type ⟨112⟩ is parallel to the rolling direction.

In the photographic method of determining the pole-figure of a metal sheet, the sheet specimen, etched to reduce its thickness so that it transmits the beam, is initially mounted perpendicular to the incident beam with the rolling direction vertical. The resulting photographs will show arcing of the Debye rings in a symmetrical arrangement about a vertical line through the centre of the film. The sheet is then rotated an angle β, say 10°, about the rolling direction and another photograph taken. The resulting pattern differs from the first one taken, as shown in Fig. 53(a) and 53(b), because the texture does not have rotational symmetry about the rolling direction, R.D. The complete pole-figure for a particular set of (*hkl*) planes can then be mapped out by taking photographs at 10° intervals and the angular extent of the arcing on the (*hkl*) Debye ring, taken from each photograph, plotted stereographically as a range of reflecting pole positions.

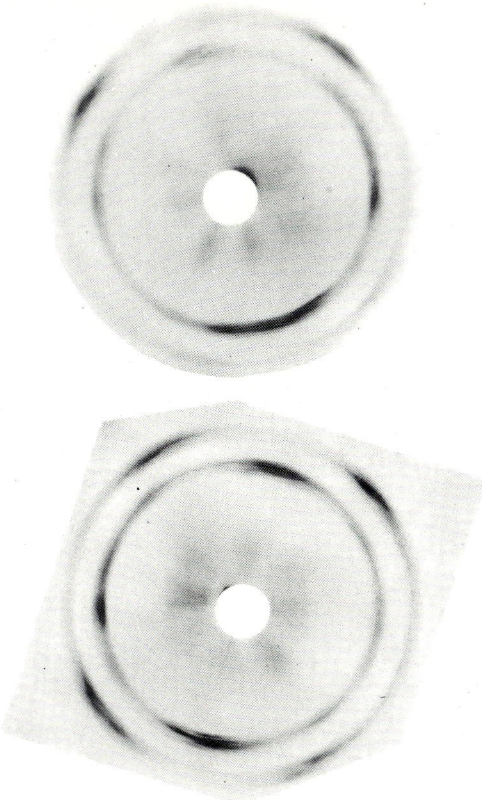

(a)

(b)

FIG. 53. Transmission X-ray photographs of aluminium sheet rolled 95% at room temperature with (a) $\beta = 0°$, (b) $\beta = 10°$.

In recent years, much of the labour in determining pole-figures has been removed by the use of automatic texture-goniometers incorporating the direct-recording of intensity with a counter. Figure 54(a) shows the principle of the transmission technique. The counter is kept fixed to receive the (*hkl*) reflection and the non-uniform distribution of intensity around the Debye ring is measured by rotating the specimen in its own plane, i.e., the δ-movement in Fig. 54(a). The specimen may also be rotated about a vertical axis, i.e., the α-movement in Fig. 54(a), so that with the combined α- and δ-movements it is possible to move the pole of the (hkl) reflecting plane over the complete surface of the pole-figure. The intensity recorded

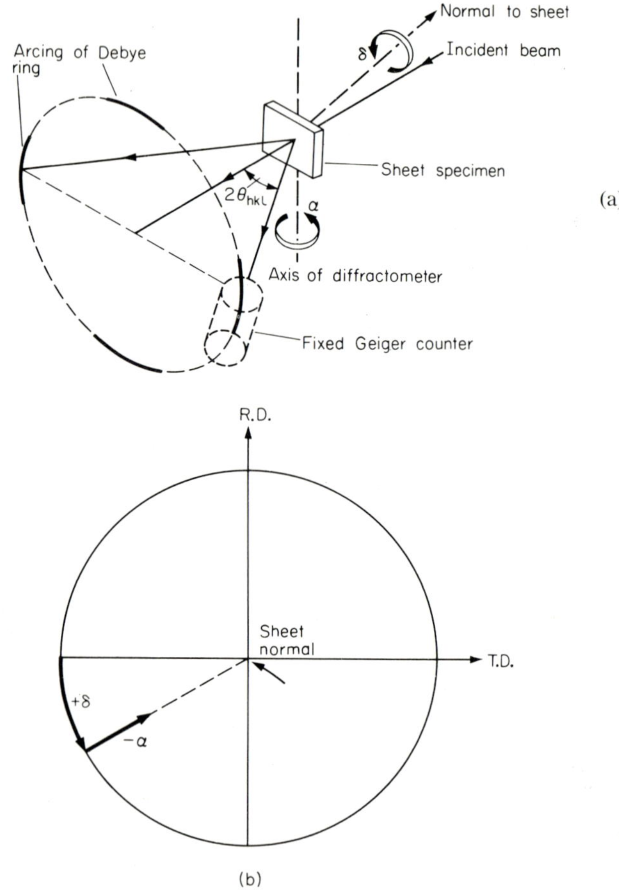

FIG. 54. (a) Schematic arrangement of the transmission counter technique for determining the rolling textures of a sheet specimen. (b) Method of plotting the intensity results on a pole-figure.

at each position of the specimen, subsequently corrected for absorption, is proportioned to the density of (*hkl*) poles at the corresponding point on the pole-figure, as shown schematically in Fig. 54(b).

In practice it is impossible to determine the central area of the pole-figure by the transmission method, because the X-ray path-length through the specimen becomes so large that absorption becomes prohibitively high. This region is usually studied with a transmission method based on that originally developed by Schulz. The principle of the method is shown schematically in Fig. 55, which also defines the angles of

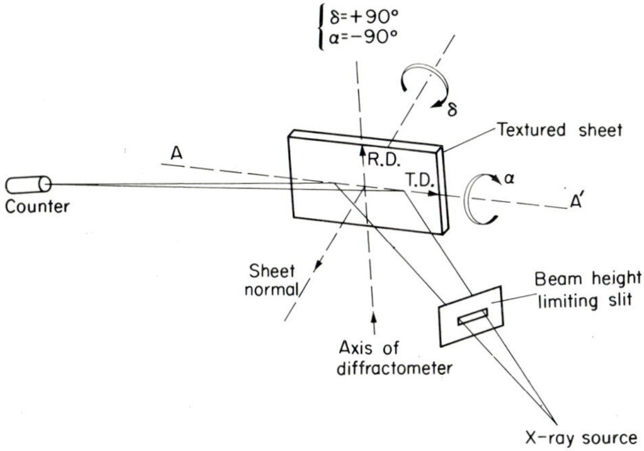

FIG. 55. Schematic arrangement of the reflection counter technique for determining the rolling texture of a sheet specimen.

rotation α and β respectively. It can be seen that the vertical height of the beam has to be limited to about $\frac{1}{2}$ mm because the α-rotation destroys the focusing condition except along the specimen axis AA'. The advantage of this technique is that no absorption correction is required for the various α- and δ-movements.

The rolling textures found in b.c.c. and c.p.h. materials may be described by the ideal orientations $\{100\}$ $\langle 011 \rangle$ and $\{0001\}$ $\langle 1000 \rangle$ respectively. F.c.c. materials exhibit a more complex texture behaviour. In Fig. 56(a) is shown a {111} pole-figure from 70–30 brass which may be described by two components of the ideal {110} ⟨112⟩ orientation. The texture of copper is reproduced in Fig. 56(b) and may be described by four equivalent orientations close to {531} ⟨112⟩. This pure metal texture has also been described in terms of a mixture of {110} ⟨112⟩ and {112} ⟨111⟩ orientations. These two textures may be characterized by an intensity ratio

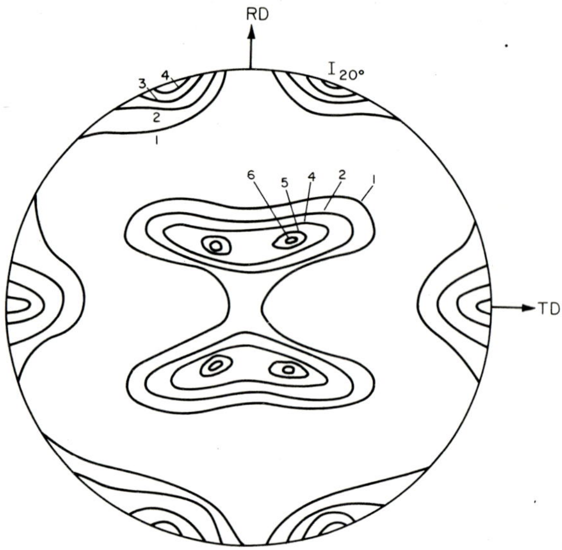

FIG. 56(a). The (111) pole-figure from 70–30 brass rolled 95% at room temperature. (After Dillamore, Smallman and Roberts, *Phil. Mag.*, 1964, fig. 1 a.)

determined from the central region of each pole-figure, but a more sensitive intensity ratio characteristic of each texture is that derived from the periphery. It can be seen from Fig. 56(a) and (b) that as the texture changes from the {110} ⟨112⟩ type to the {531} ⟨112⟩ type, the intensity peak on the "equator" of the pole-figure moves towards the centre, whilst that at 20° to R.D. remains stationary within the scatter of the ideal orientation. A convenient measure of the proportion of (110) [$\bar{1}$12] orientation present at any time may thus be obtained from the ratio of intensities of {111} poles at the transverse direction (I_{TD}) to that 20° from the rolling direction (I_{20}).

Recent work on the deformation texture of f.c.c. metals and alloys has shown that a correlation exists between the deformation texture and the stacking fault energy. For example,

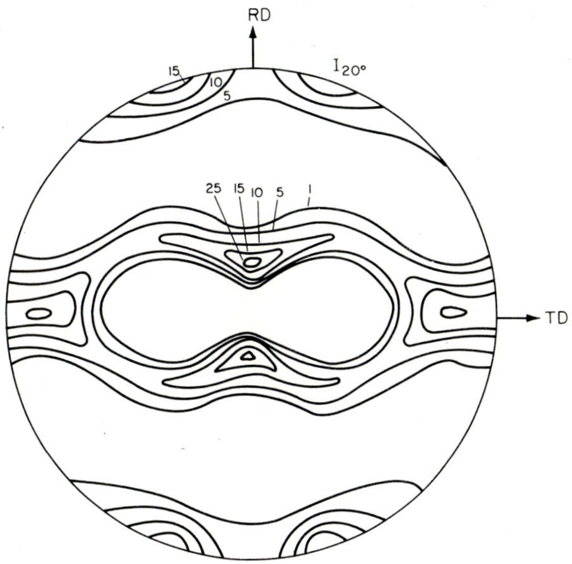

FIG. 56(b). The (111) pole-figure from copper rolled 95% at room temperature. (After Dillamore, Smallman and Roberts, *Phil. Mag.*, 1964, fig. 1a.)

the variation of rolling texture with composition in those alloy systems for which the stacking fault energy has been determined, show that the {110} ⟨112⟩ texture (Fig. 56a) is exhibited by copper, silver and gold alloys when $\gamma/\mu b$ is less than about 4, where γ is the stacking fault energy of the materials, μ the shear modulus and b the Burgers vector.

The stacking fault energy parameter is important in governing the texture behaviour because of its influence on the cross-slip process. From the rotations for crystals of various orientation, it can be demonstrated that when extensive cross-slip occurs the "alloy type" rolling texture $\{110\}$ $\langle 112 \rangle$ transforms into the "pure-metal" type texture. Using a single crystal analogue of the process it may be concluded that the (110) $[\bar{1}12]$ orientation is broken up at some point in the deformation process after the onset of stage III work hardening. The later the stage in the deformation process at which cross-slip becomes sufficiently extensive to affect the orientation, the greater will be the amount of material with the (110) $[\bar{1}12]$ orientation. If a measure of the amount of (110) $[\bar{1}12]$ orientation present in the textured material could be obtained, this would provide an indication as to the stage reached in the deformation process, and whether cross-slip had become sufficiently extensive to affect the orientation.

From a study of the texture exhibited by a given metal it should therefore be possible to estimate the stacking fault energy. Such a method of estimating the stacking fault energy is related to the method proposed by Seeger and his co-workers for determining γ from the onset of the third stage of the single crystal stress–strain curves in that the parameter obtained by measuring the pole-figure is related to the onset of stage III in the single crystal stress–strain curve. The method differs, however, in that no arbitrary cross-slip mechanism is assumed, and, moreover, it is essentially the value of strain corresponding to cross-slip operation which is measured, not the stress value. The validity is dependent on the assumption that the single crystal stress–strain curves of metals of the same value of $\gamma/\mu b$ with equivalent thermal activation and for the same orientation are identical when the stress is measured as a fraction of the shear modulus. It seems likely that this is a reasonable assumption. Values of the stacking-fault energy estimated from rolling texture determinations for several f.c.c. metals are listed in Table 2.

TABLE 2. VALUES OF STACKING FAULT ENERGY DERIVED FROM TEXTURE MEASUREMENTS

(After Dillamore and Smallman. *Phil. Mag.*, **12,** 191, 1965.)

Metal	Cu	Ag	Au	Ni	Pt	Pd	Rh	Th	Yb	Ce
γ ergs/cm^2	85	21	52	450	95	180	750	115	<10	<5

SMALL-ANGLE X-RAY SCATTERING

The scattering of X-ray intensity into the low angle region ($\epsilon = 2\theta < 10°$) arises from the presence of inhomogeneities within the material (such as small clusters of solute atoms) having dimensions only 10 to 100 times the wavelength of the incident radiation. The origin of the scattering can be attributed to the differences in electron density between neighbouring heterogeneous regions and to the differences in electron density between the heterogeneous regions and the surrounding matrix, so that precipitated particles afford the most common source of scattering; other heterogeneities such as dislocations, vacancies and cavities must also give rise to some small angle scattering, but the intensity of the scattered beam will be much weaker than that from precipitated particles. The intensity scattered at small angles by a sample containing N scattering aggregates is given by the approximate formula

$$I = Nn^2 I_e \exp\left[-4\pi^2\epsilon^2 R^2/3\lambda^2\right]$$

where n represents the difference in number of electrons between the particle and an equal volume of the surrounding matrix, R is the radius of gyration of the particle, I_e is the intensity scattered by an electron, ϵ is the angle of scattering and λ is the wavelength of the X-rays. From this equation it can be seen that the intensity of small angle scattering is zero if the inhomogeneity, or cluster, has a density equivalent to that of the surrounding matrix even if it has quite a different crystal structure. On a plot of $\log_{10} I$ as a function of ϵ^2, the slope near the origin, $\epsilon = 0$, is given by $-(4\epsilon^2/3\lambda^2)R^2 \log_{10}e$ and for Cu$K\alpha$ radiation the radius of gyration of the scattering aggregate R in Angstrom units is approximately $\frac{2}{3}\sqrt{\text{slope}}$.

The technique has been used to study the early stages of precipitation and the aggregation of point defects.

MICRO-PROBE ANALYSER

This instrument is used for identifying and quantitatively analysing elements in regions of a specimen surface as small as 1μ in diameter with a detection sensitivity of about 10^{-14} grams. An ordinary metallographic specimen is irradiated with a finely focused beam of electrons and the resulting emission of characteristic X-rays for each element excited in the sample is then recorded with a Geiger counter (usually of the proportional type) and then analysed. This technique, which is largely non-destructive, is known as X-ray microanalysis.

The principle of operation is illustrated schematically in Fig. 57. A tungsten filament forming part of a triode electron gun is heated *in vacuo* to produce electrons which are focused by magnetic lenses into a spot less than 1μ in diameter on the surface of the specimen. This is mounted on a table which can be rotated to bring the specimen under an optical microscope. Scanning coils deflect the finely-focused spot so that it scans a raster approximately $\frac{1}{2}$ mm square at the lowest magnification,

Some of the electrons scanning the sample are reflected with little loss of energy while others penetrate the surface and excite the emission of X-rays. A proportion of the X-rays emitted is allowed to emerge from the column through an exit port inclined at an angle of 20° to the surface of the specimen. They then strike a curved analysing crystal usually of lithium-fluoride in the X-ray spectrometer. This crystal can be rotated relative to the incident X-ray beam and set at the correct Bragg angle for reflecting the characteristic emission of a chosen element through the window of a proportional counter. The signal

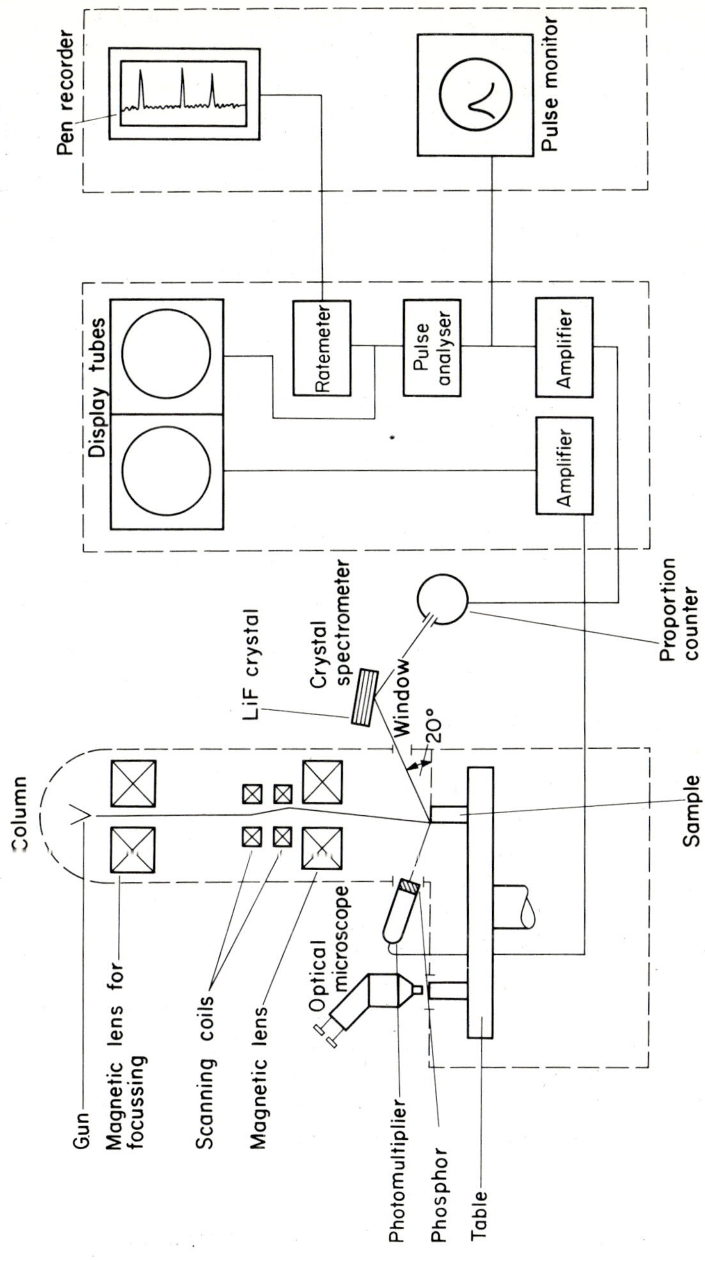

FIG. 57. Schematic diagram of micro-probe analysis equipment. (*Courtesy of the Cambridge Instrument Co.*)

from the counter is amplified and used to modulate the intensity of a beam scanning the long persistence screen of one of two similar cathode ray tubes.

The beam scanning the other cathode ray tube is similarly modulated by an amplified signal from a photomultiplier attached to one end of a plastic light guide. A phosphor energized by being struck by some of the high energy reflected electrons is attached to the other end near the surface of the sample. The beam of electrons scanning the sample and the beams scanning the cathode ray tubes are synchronized so that there is an exact correlation between the area scanned on the surface of the specimen and the X-ray and electron images. The electron image is recognizably similar to an optical image of the area scanned while the X-ray image shows the distribution and approximate concentration of a selected element over the same area.

For quantitative analysis, the monochromatic X-ray signal is passed through a ratemeter which counts the number of pulses detected each second and generates a proportional d.c. voltage suitable for operating a sensitive pen recorder. If the crystal is slowly rotated through its range of Bragg angles, the pen recorder will plot, in turn, peaks corresponding to the intensities of the characteristic lines emitted by each element present in the sample. For quantitative determinations, these peaks are then compared with similar peaks produced by pure standard samples of each element. The instrument will give averaged quantitative results over an area up to $\frac{1}{2}$ mm square if the scanning facilities are used. An exact analysis of any area approximately 1 μ in diameter and $\frac{1}{2}\mu$ deep can be made if the scan is stopped and the electron beam used as a stationary probe.

In an alternative arrangement the X-rays may by-pass the crystal and pass directly into the counter for analysis by pulse height discrimination since the energy of the pulse from the counter varies with the X-ray wavelength. A 35 mm camera is provided for photographing the images shown in the tubes for later examination and record purposes.

The technique is useful for investigating the problems concerned with small scale variations in alloy composition such as encountered in studies of precipitates, inclusions, segregation of alloying elements, diffusion gradients, and grain boundary de-nudation or enrichment. It is also useful in studying corrosion effects, particularly oxide scale, etc., and fracture surfaces. The main limitation of the technique is that, at present, it is impossible to detect the soft X-rays from elements of atomic numbers lower than 12.

THE OBSERVATION OF DISLOCATIONS BY X-RAY DIFFRACTION CONTRAST

With the X-ray diffraction techniques already outlined it is possible to determine the structure and lattice parameter of a metal or alloy as well as its composition, orientation and degree of perfection. Nowadays it is also possible to study individual crystal defects with X-rays by detecting differences between the intensity *diffracted* by regions of the crystal near such defects (e.g., dislocations) and more nearly perfect regions of the crystal. Two different methods are shown schematically in Fig. 58, both of which use collimated monochromatic $K\alpha$-radiation and photographic recording. Figure 58(a) shows the back-reflection method in which the beam is reflected from the surface of the crystal set at the Bragg-angle for a particular

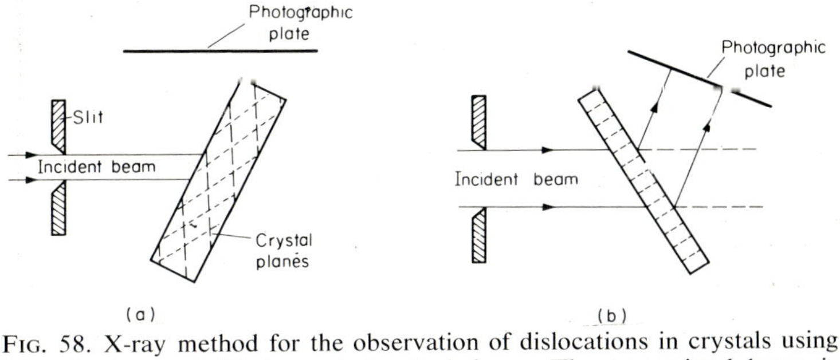

FIG. 58. X-ray method for the observation of dislocations in crystals using (a) reflection and (b) transmission techniques. The transmitted beam is intercepted in both cases.

set of lattice planes; Figure 58(b) shows the thin-crystal transmission method. Any imperfections give rise to local changes in diffracted or transmitted X-ray intensities and, consequently, dislocations show up as bands of contrast, some 5 to 50μ width. No magnification is used in recording the diffraction image, but subsequent magnification of up to 500 times may be achieved with high resolution X-ray emulsions. Large areas of the crystal to thicknesses of 10–100μ can be mapped using scanning techniques provided the dislocation density is not too high ($\ngtr 10^6$ cm^{-2}).

The X-ray method of detecting lattice defects suffers from the general limitations that the resolution is low and exposure times are long (12 hours). By comparison, the thin-film electron microscopy method (see next chapter) is capable of revealing dislocations with a much higher resolution because the dislocation image width is only 100 Å and magnifications up to 100,000 times are possible. The X-ray method does, however, have the great advantage of being able to reveal dislocations in crystals which are comparatively thick (~0·1 cm, cf. ~1000 Å in foils suitable for transmission electron microscopy). The technique has been used for studying in detail the nature of dislocations in thick single crystals with very low dislocation densities, such as found in semi-conducting materials; the back-reflection and transmission methods are suitable for studying the dislocation structures of surface regions and the interiors of crystals respectively. Figure 59 shows an example of an X-ray topograph revealing dislocations in aluminium by this technique.

Diffraction methods using either X-rays or electrons can reveal individual dislocations, stacking-faults, sub-boundaries, grain-boundaries and precipitates. In all cases, the defects are observable only when they produce an atomic displacement with a component normal to the set of reflecting planes responsible for the diffraction contrast. In principle, this condition allows the displacement vector of a stacking-fault or Burgers vector of a dislocation to be determined from observations of

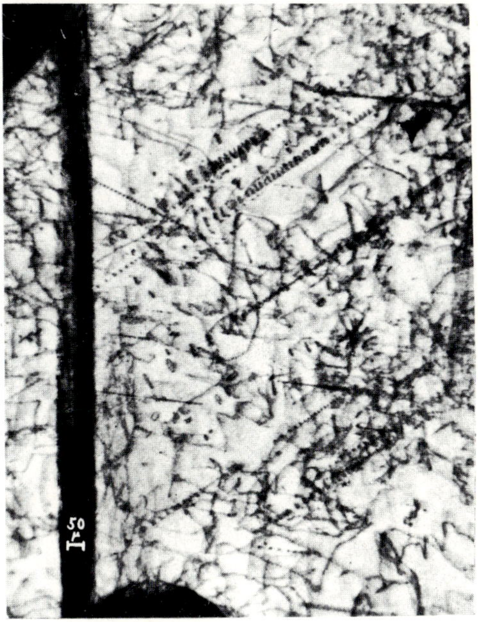

FIG. 59. X-ray topograph of aluminium single crystal showing helical dislocations, rows of dislocation loops in a ⟨110⟩ and a twin boundary on the left. (After A. Lang.) Magnification × 40.

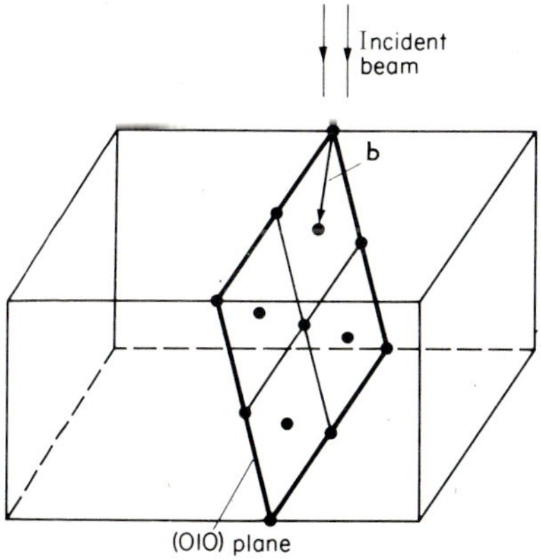

FIG. 60. Schematic diagram to show the conditions for which a dislocation with vector **b** would be "invisible"

the contrast of the image formed by different sets of lattice reflecting planes. For example, in an f.c.c. crystal a dislocation with $\mathbf{b} = \frac{a}{2}\left[101\right]$ would be "invisible" if the 020 reflection is used to produce the image, as shown in Fig. 60. This technique has been applied with considerable success in electron microscopy and is described in detail in Chapter IX.

SUGGESTIONS FOR FURTHER READING

Direct Observations of Imperfections in Crystals. Interscience Publications, 1963.

B. D. Cullity, *The Elements of X-ray Diffraction*, Addison-Wesley, 1959.

V. E. Cosslett, X-ray microscopy and microanalysis, *Met. Reviews*, **5**, 225 (1960).

Symposium on Electron Probe Microanalysis, *J. Inst. Metals*, **91** (1962–63).

CHAPTER VII

ELECTRON MICROSCOPY – I
ELECTRON MICROSCOPE

ELECTRON WAVELENGTH

It was discussed in Chapter I that in order to increase the resolving power of an optical microscope it is necessary to employ light of shorter wavelengths. Thus, ultra-violet light may be used, but even then the resolution can hardly exceed 1000Å. As a consequence, the electron microscope has been developed to allow the resolution of structure having dimensions down to 10Å.

Moving electrons may be considered as particle waves with a wavelength λ given by the De Broglie equation

$$\lambda = h/mv$$

where h is Planck's constant and mv is the momentum of the electron. If the electrons are accelerated by voltage V, the kinetic energy they acquire is

$$mv^2/2 = Ve$$

where e is the charge on an electron and therefore

$$\lambda = h/\sqrt{2mVe} = \sqrt{150/V}\ \text{Å}$$

When the velocity of the electrons is comparable to the speed of light, c, a relativistic correction to this equation is required and the expression becomes

$$\lambda = \sqrt{150/(1 + 97{\cdot}88 \times 10^{-8}V)} \simeq \sqrt{150/(1 + 10^{-6}V)}.$$

The relativistic correction to the classical expression for λ only amounts to about 5% at 100 kV; at 10^6 kV there would be an error of $1/\sqrt{2}$, i.e., 30%. The values of λ at the accelerating voltages commonly used in electron microscopy are shown in Table 3, from which it is seen that the variation of λ with accelerating voltage is small.

TABLE 3. VARIATION OF ELECTRON WAVELENGTH WITH APPLIED VOLTAGE

Accelerating voltage (kV)	Electron wavelength (Å)
70	0·0449
80	0·0418
90	0·0392
100	0·0370

THE ELECTRON MICROSCOPE

General Description

An electron microscope consists of an electron gun and an assembly of lenses all enclosed in a column evacuated to about 10^{-5} mm Hg. Figure 61(a) shows a schematic arrangement of such an instrument and Fig. 61(b) a vertical cross-section through an A.E.I. E.M.6 model. The optical arrangement is similar to that of a light microscope but additional stages of magnification are used. The lenses usually are of the magnetic type in which current-carrying coils are completely surrounded by a soft iron case except for a narrow gap in the bore, about 1 cm diameter. Apertures are placed below the condenser, objective and first projector lenses to limit the divergence of the electron beam. The use of apertures is discussed in detail in the section on selected area diffraction. The lenses are

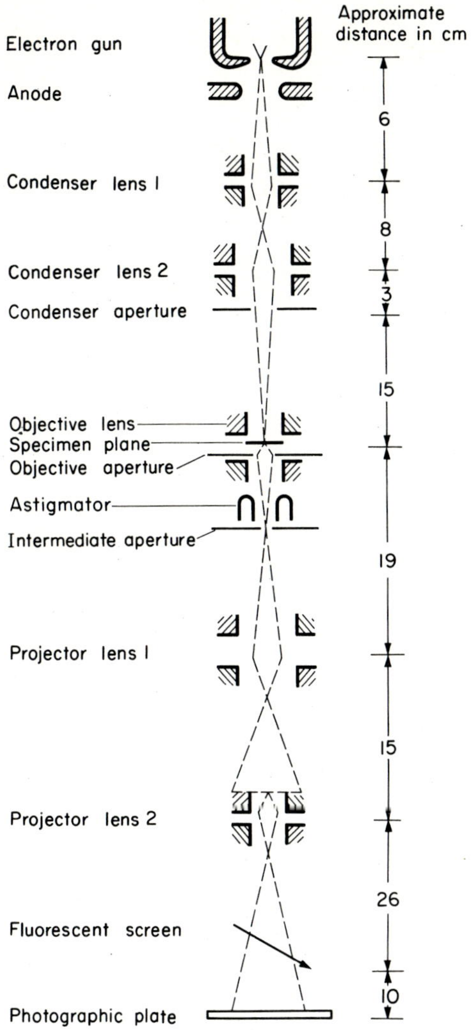

FIG. 61(a). Schematic arrangement of E.M.6 column

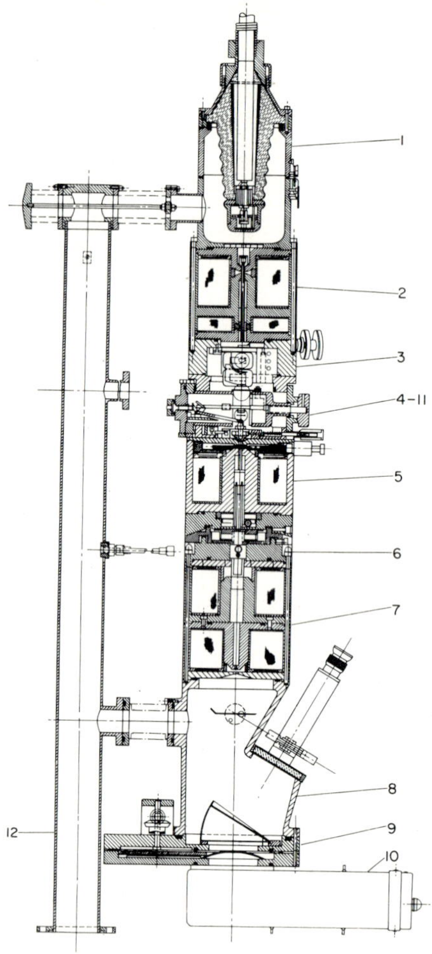

FIG. 61(b). Cross-section through E.M.6 microscope. (*Courtesy of A.E.I.*)

1 – 100 kV electron gun
2 – two stage condenser lens system
3 – beam deflector stage
4 – specimen chamber
5 – objective lens
6 – intermediate alignment stage
7 – two-stage projector lens system
8 – viewing chamber and screen
9 – vacuum valve to seal-off the column from the camera
10 – plate camera
11 – specimen manipulator
12 – high vacuum manifold

energized by d.c. and their focal lengths can be varied by altering the currents passing through the coils. This method of altering the focal lengths is more satisfactory than physically adjusting the positions of the lenses as is done in the case of glass lenses in optical microscopy. The two condenser lenses collimate the electron beam from the gun towards the object which is placed in the bore of the objective lens. The objective lens produces a magnified image of the specimen in the object plane of the intermediate, or first projector lens, and a small part of this image is magnified to produce a second image in the object plane of the second projector lens. Magnification by the second projection lens produces a final image which may be viewed on a fluorescent screen or recorded on 35 mm film or photographic plate.

THE ELECTRON GUN AND CONDENSER LENS SYSTEM

The function of the electron gun is to supply a steady stream of electrons; to do this the filament is operated in a saturated condition. In this state the excitation current is set at a minimum value, above which the number of electrons emitted by the filament remains constant. When the accelerating voltage is 100 kV the filament operates at a negative potential of 100 kV, the anode plate at earth potential and the filament shield at a negative potential of 98 kV. This arrangement allows the maximum electron beam intensity, consistent with a small source size. The difference in potential between the filament and shield is known as the gun bias, and governs the effective aperture for the passage of electrons.

Under these conditions the electrons are accelerated into the lens system of the microscope as a diverging beam which is focused by the condenser lenses to form an illuminated spot on the specimen. At any magnification, the brightest possible image is formed when the incident beam illuminates only the area of specimen whose image appears on the screen. The area of specimen used to form the final image decreases with increasing magnification and, to maintain maximum image brightness, provision must be made for decreasing the spot size. To see how a variable spot size is obtained it is necessary to consider the two stage condenser lens system which is standard in most modern microscopes. A schematic diagram of the electron beam path through the lens systems of the E.M.6 is shown in Fig. 62. The first lens C_1 produces a greatly demagnified image

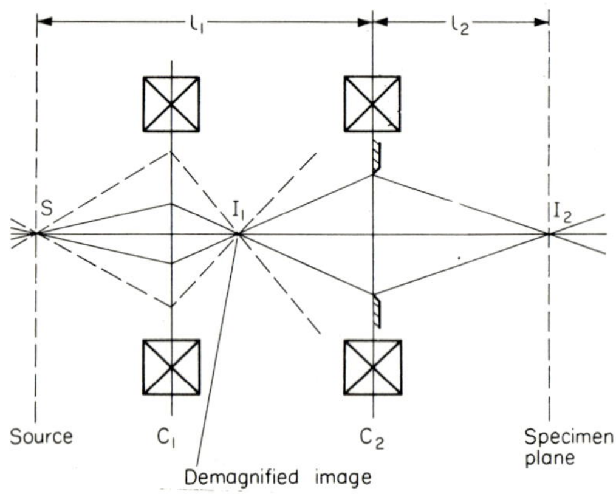

FIG. 62. The double-condenser lens system, (After Haine and Cosslett, *The Electron Microscope*, p.165, fig. 8.5, E. Spoon Ltd., 1961.)

of the source S_1 at I_1, while the second lens magnifies the image I_1 to form the spot on the specimen at I_2. Demagnification of the spot may be continuously varied over a wide range by altering the power of the first lens and a very small spot may be produced.

Two advantages from the designers point of view result from the two condenser system shown in Fig. 62. First it provides a long working space I_2, in which the specimen airlock and beam deflector stage can be inserted, and secondly because of the sharpness of the illumination, the general contamination of surrounding areas of the specimen is minimized.

SPECIMEN ASSEMBLY

A typical specimen holder, grid and thinned metal specimen is shown in Fig. 63(a). A small piece of metal foil is placed between two grids and then into the holder. The specimen

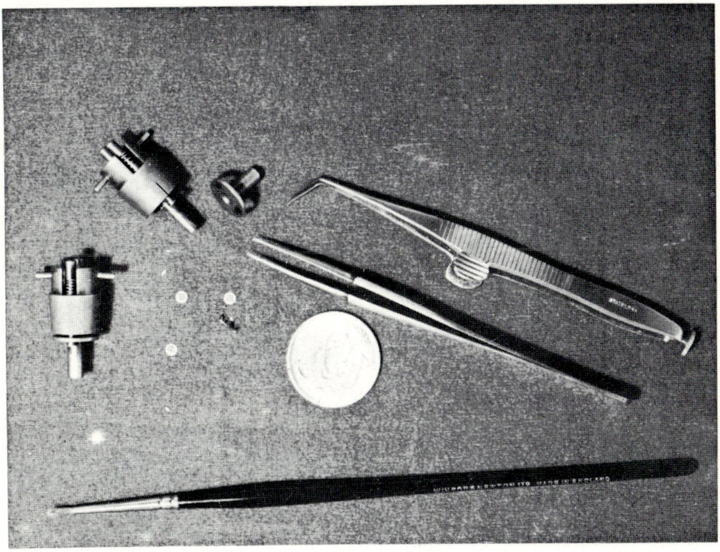

FIG. 63(a). Photograph showing thinned metal foil, specimen grid and specimen holder.

chamber, shown in Fig. 63(b) consists of an airlock, to facilitate insertion and removal of the specimen holder into the evacuated column, and a specimen stage to hold the specimen in position. The specimen stage is movable in all directions in a horizontal plane to allow scanning of the sample and for most metallurgical applications has one or more horizontal axes about which the specimen may be tilted to vary the contrast at defects such as dislocations and precipitates. During all lateral and rotational movements, the area of specimen under examination must, of course, always remain in the back focal plane of the combined condenser lenses. The important exception to this rule is in the study of magnetic or superconducting domain structures when the specimen must not lie in a magnetic field, e.g., that of the objective lens.

LENS DEFECTS

The magnetic lenses of an electron microscope suffer from the same defects as glass lenses in an optical microscope, namely chromatic aberration, spherical aberration and astigmatism. Chromatic aberration is the spread of electrons in the imaging plane and results from the focusing of electrons having different velocities, and from fluctuations in focal

FIG. 63(b). Specimen holder emerging from the air-lock assembly to be seated in the tilting block.

length. The two sources of chromatic aberration are reduced by stablizing the high voltage supply and lens current respectively. Spherical aberration is the defect which results from the outer regions of a lens forming images at distances further from the lens than the images formed by the inner regions. The amount of spherical aberration is proportional to α^3, where α is the semi-apex angle subtended by the lens at the object. Spherical aberration can be reduced by imposing strict limitations on the geometry of the lens. Astigmatism is the most important defect in a magnetic lens, since only extremely small mechanical or magnetic asymmetry is required to significantly deviate the paths of 100 kV electrons. Although lens design and accurate machining of the pole pieces minimize astigmatism it is necessary, for the condenser and objective lenses at least, to provide an external correction. This is accomplished by applying a magnetic field, the magnitude and direction of which can be varied to distort the lens field in such a way as to restore symmetry.

CONTROL OF FOCUSING AND MAGNIFICATION

The focal length of a lens is proportional to v, the velocity of the electrons, which is controlled by the voltage on the electron gun, and inversely proportional to the square of H, the axial component of the magnetic field strength of the lens. This second component depends on the current passing through the lens coil and is the variable used to focus finely the lens during operation; by this means, the focal length can be changed from a few centimetres to a few millimetres.

In normal metallurgical practice, the focal length of the objective lens is so adjusted that an image is formed in the same plane as the intermediate aperture (see Selected Area Diffraction, on p. 119). Consequently, the contribution by the objective lens to the final magnification is fixed and the variation in magnification is achieved with the two projector lenses P_1 and P_2. For a given image distance, the magnification produced by a lens increases with decreasing focal length.* Consequently, the maximum magnification is obtained when P_1 and P_2 operate at minimum focal lengths. In practice, P_2 is often operated at a constant very small focal length and control of the magnification (and focusing of the diffraction pattern on the screen, see p. 120) is effected by varying the

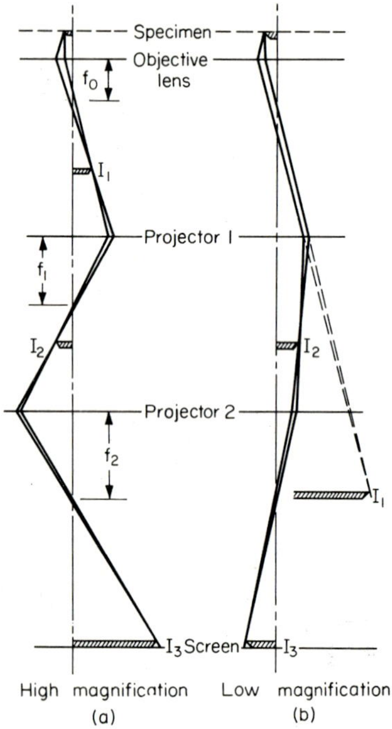

FIG. 64. Ray diagram showing how the final image is formed in the electron microscope at (a) high and (b) low magnifications.

*From the lens formula $\frac{1}{f} = \frac{1}{u} + \frac{1}{v}$, where f is the focal length of the lens, and u and v are the object and image distances respectively, it follows that the magnification $m = \frac{v}{u} = \frac{v-f}{f}$.

excitation current of the projector or intermediate lens P_1.

For operation at high magnification, i.e., ×20,000 to about ×120,000, all three imaging lenses (objective, projector 1, and projector 2) form inverted images denoted I_1, I_2, and I_3 in Fig. 64(a). The maximum magnification is limited by the minimum focal lengths of these lenses, and this minimum is limited by the degree of "pincushion" distortion, which can be tolerated at the edge of the field of view.

For operation at low magnification, i.e., ×1500 to about ×16,000, the intermediate lens, P_1, is used at long focal length. In this case, an erect image I_2 of I_1 is formed in the back focal plane of P_2 as shown in Fig. 64(b). The final image I_3 is inverted and small. Projector 2 works at fixed focal length and the minimum magnification is limited by the degree of "barrel" distortion tolerated at the edge of the field.

BRIGHT-AND DARK-FIELD IMAGES

Of the electrons emerging from a thin metal foil, some have been diffracted by the lattice planes and the rest simply undeviated, i.e., transmitted. If the diffracted rays are allowed to contribute to the final image, the quality will be poor because of the lens aberrations discussed on p. 114. (The diffracted rays are gathered by the outer regions of the objective lens.) In the technique known as thin film transmission microscopy, either the transmitted beam only or one of the diffracted beams only is allowed to contribute to the final image. This is achieved by inserting a small aperture in the objective lens, see Figs. 61(b) and 65, to block either the transmitted or diffracted rays.* When the image is formed by the beam which is transmitted directly through the crystal, diffracted beams being blocked, a bright-field picture is obtained. When the image is formed by one of the diffracted beams, the direct beam being blocked, a dark-field

*An objective aperture of 40 μ is equivalent to a semi-angular aperture of somewhat less than 10^{-2} radians and since $\lambda \sim 0{\cdot}04$Å then $2\theta > 10^{-2}$ radians. Hence beams with Bragg reflections for which $2\theta > 10^{-2}$ radians will be blocked by the aperture which, for metals, means all diffracted beams.

picture is obtained. In practice, the diffracted ray for dark-field microscopy is selected by switching to diffraction conditions, moving the objective aperture until it surrounds only the spot of the reflection being studied and then switching back to normal microscope conditions. Bright-field illumination is used more frequently than dark field illumination, because the dark field image is distorted since it is formed by rays travelling at an angle to the instrument axis and therefore suffers from spherical aberration and astigmatism. However, in some instruments it is possible to tilt the electron gun, or incident beam, such that the selected diffracted beam travels down the axis and, as a result, a perfect dark-field image can be obtained. The general arrangement of the electron beam relative to the objective aperture for bright and dark-field operation is shown in Fig. 65. Examples of micrographs taken under bright and dark-field conditions are shown in Figs. 97 and 106.

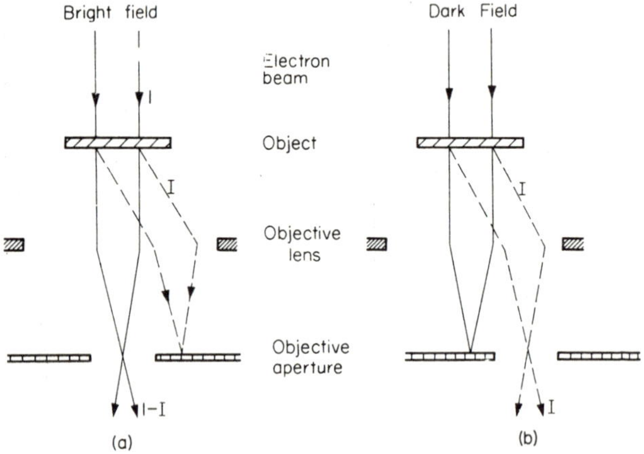

FIG. 65. Ray diagrams to illustrate (a) bright-field and (b) dark-field microscopy. The diffraction pattern is formed in the plane of the objective aperture and in the bright-field image the aperture prevents the diffracted rays from contributing to the image.

SELECTED AREA DIFFRACTION

The most important part of the technique of transmission electron microscopy is selected area diffraction. Only from S.A.D. is it possible to obtain information concerning the crystallography of the specimen and the electron diffraction conditions pertaining to any small area (~ 0·5 μ diameter) selected from the specimen. The present section will be confined to the formation of a selected area diffraction pattern and its relationship to the transmission electron micrograph. Chapter IX will deal with the indexing and interpretation of the pattern obtained by S.A.D.

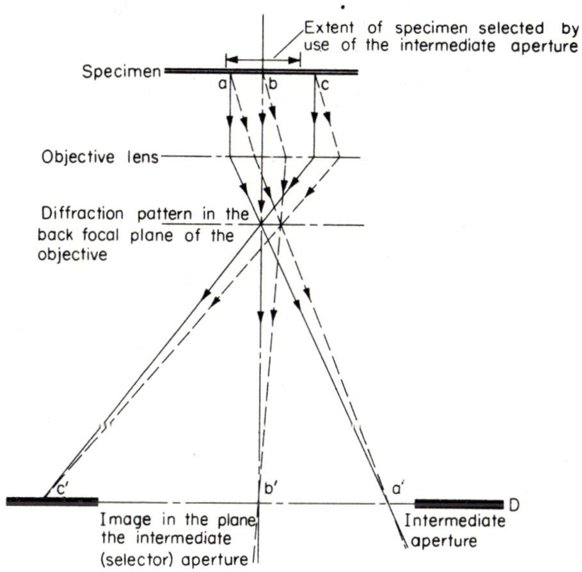

FIG. 66(a). Schematic diagram showing the position of the first image and the diffraction pattern formed by the objective lens. (After Halliday and Phillips from *Techniques for Electron Microscopy*. fig. 12–21, edited by D. Kay, Blackwell, 1961.)

The objective lens produces an intermediate image of the specimen in front of the intermediate or first projector lens as shown in Fig. 66(a) at $a'b'c'$. Electrons diffracted, by a given lattice plane in the specimen, are focused to a point in the rear focal plane of the objective lens. The distance of this

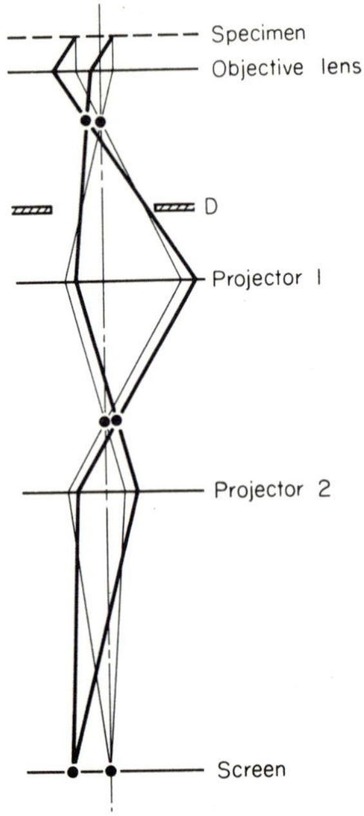

FIG. 66(b). The complete ray diagram for the formation of a diffraction pattern at the screen (*A.E.L. Handbook*).

point from the centre axis of the microscope is proportional to θ, the Bragg angle for the diffracting plane (see p. 66). In general, many planes will be suitably oriented for diffraction and a pattern of focused spots will be formed. For the limiting case of a random polycrystal, hundreds of planes produce diffraction and a pattern of diffraction rings, not individual spots, results.

Referring to Fig. 66, it can be seen that the transmitted and diffracted electrons from the same area of specimen, e.g., *a b c*, meet in the plane of the intermediate image. Consequently, a small aperture located in this plane permits the examination of the diffraction pattern from an area of intermediate image exactly equal to the area of the aperture. This is known as selected area diffraction. While the transmission picture is on the screen, the intermediate or S.A.D. aperture is inserted into the plane of the intermediate image and, by moving the aperture in this plane, the area being studied is selected when it is encircled by the image of the aperture as seen on the screen.

There are four sources of error to be considered when taking selected area diffraction patterns. These are spherical aberrations due to the lenses, incorrect focusing of the objective lens, incorrect focusing of the intermediate lens and on the Siemens Elmiskop I microscope, the magnetic coupling between the objective and intermediate lenses. The first two sources provide the largest errors and will now be discussed in more detail.

It was stated on p. 115 that the degree of spherical aberration increases as we move radially away from the centre of a lens. This means that spherical aberration will exert a larger effect on the positions of high order reflections than those of lower order (N.B., the further a spot is away from the centre of a diffraction pattern, the higher its order.) However, since it is difficult to calibrate the effects of spherical aberration on a selected area diffraction pattern, it is more convenient to minimize the defect and to use the smallest intermediate aperture which will give satisfactory results ($\sim 5\ \mu$).

From the point of view of the operator the main practical difficulty in S.A.D. is to ensure that the intermediate image is formed in the plane of the intermediate aperture. Consider Figs. 67 (a) and (b) which show the relationship between the selected area diffraction pattern and the intermediate image for correctly focused and underfocused objectives respectively.

For the correctly focused objective the intermediate aperture allows all the transmitted and diffracted electrons from the selected area to pass through it. In the case of the underfocused objective, however, the aperture obstructs some of the electrons transmitted by the area corresponding to the S.A.D.

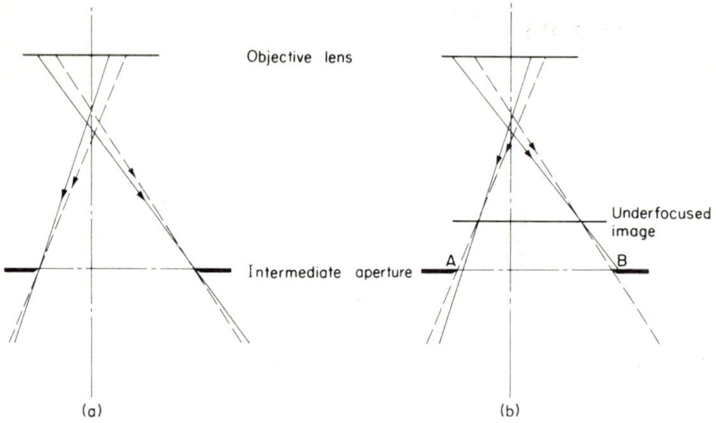

FIG. 67. The paths of transmitted rays (continuous lines) and diffracted rays (broken lines) (a) for a correctly focused objective and (b) for an underfocused objective.

pattern (e.g., at *B* in Fig. 67b) and allows others from outside this area to pass through it (at *A* in Fig. 67b). In other words, the selected area is laterally displaced from the area producing the diffraction pattern. A similar displacement, in the opposite sense, occurs if the objective lens is overfocused.

In order to focus the objective lens, use is made of the fact that, under correct S.A.D. conditions, the intermediate image and intermediate aperture are co-planar. This means that, as far as the projector lenses are concerned, the intermediate image and the intermediate aperture behave as objects located at exactly the same place. Consequently, the field of view and the edge of the aperture will be simultaneously in sharp focus on the screen. The procedure to be adopted is as follows:

1. With the transmission picture on the screen, the intermediate (diffraction) aperture is inserted.
2. Adjust the current in P_1 the projector (intermediate) lens until the image of the inner annular edge of the aperture is sharply in focus. As a general rule, this can be done at one and only one setting for P_1, i.e., at only one magnification. This so called S.A.D. magnification is usually about ×20,000. The exception to the rule is the Philips E.M. 200 which has two intermediate lenses. In this case, the intermediate image and the diffraction aperture are in focus on the screen for many different settings of the two lenses, i.e., at several magnifications. Consequently, the E.M. 200 can be used to take S.A.D. at almost any magnification.
3. That part of the image of the specimen seen inside the aperture is then focused by adjusting the objective lens current. The objective lens is then correctly focused for S.A.D.
4. By adjusting its alignment screws, the intermediate aperture is traversed across the intermediate image until its image on the screen covers the area being studied.
5. The selected area diffraction pattern may be seen on the screen by removing the objective aperture and reducing the strength of the intermediate lens, P_1, until it is focused on the diffraction pattern which is formed in the back focal plane of the objective (see Fig. 66). This pattern is then magnified and projected onto the screen, as shown in Fig. 66(b), with projector P_2 working at fixed focal length. The pattern may be sharpened by underfocusing the second condenser.

The S.A.D. pattern and the transmission image are subject to two sources of rotation about the microscope axis. The first is a 180° rotation arising from the difference in sense of the intermediate image and the diffraction pattern as shown in Fig. 66(a). The second source of relative rotation arises from the fact that to produce an electron diffraction pattern the magnitude of the current energizing the intermediate lens *P* is different from that necessary to produce a micrograph. There is therefore a relative rotation between the pattern and micrograph and this must be known before crystallographic directions are transferred from the diffraction pattern to the micrograph. The rotation at each magnification, i.e., at each setting of the first projector lens, may be calibrated by photographing a thin specimen containing a lattice feature for which the orientation is known. This known orientation on each plate is then compared with the S.A.D. pattern of the lattice feature, and the angle of rotation measured (see Fig. 68). Thin needle-shaped crystals of molybdenum tri-oxide are usually used for this purpose since these are easily deposited on the crucible lid when ammonium molybdate is heated in a closed crucible;

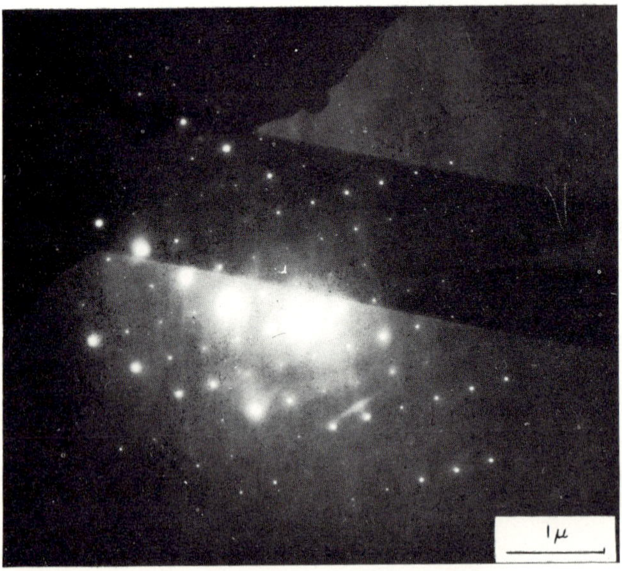

FIG. 68. Superimposed micrograph and diffraction pattern from a MoO_3 crystal to show the relative rotation between the ⟨100⟩ edge of the crystal and the ⟨100⟩ direction of the diffraction pattern.

the edges of the crystal are parallel to ⟨100⟩ directions. It should be noted that the above calibration is not the same for all accelerating voltages. This is because of a third image rotation which is proportional to $1/v$ where v is the velocity of the electrons (v is proportional to the accelerating voltage).

RESOLUTION

The resolution of any microscope is limited, by errors arising from diffraction, to $\delta = 0{\cdot}5\ \lambda/\mu \sin \alpha$ (see p. 8), where λ is the wavelength of the radiation used, μ is the refractive index of the medium between the specimen and the objective, and α is the half angle subtended by the objective at the specimen. In the case of an electron microscope, however, spherical aberration also imposes a limit on the resolution, and, since the two limits arise independently of each other, the larger determines the resolution. Consequently, the designer imposes a limit on the spherical aberration error equal to the diffraction error. i.e.,

$$C_s \alpha^3 = 0{\cdot}5\ \lambda/\mu \sin \alpha$$

where C_s is the spherical aberration constant and, because it is a small angle, $\alpha \simeq \sin \alpha$ and the optimum value of α is given by

$$\alpha = 0{\cdot}85\ [\lambda/C_s]\ 1/4$$

For a microscope operating at 100 kV, $\lambda \simeq 0{\cdot}04$ Å, $C_s \simeq 0{\cdot}05$ cm, hence $\alpha \cong 10^{-2}$ radians (cf. an optical microscope for which α can be as large as 70°), and the limit of resolution is about 4 Å. In general, other instrumental errors such as astigmatism raise the limit to about 7 Å. It therefore follows that for metal crystals, for which the interplanar spacings are much less than 7Å, direct imaging of the lattice planes is not possible.

RECORDING THE IMAGE

In the electron microscope, the depth of field D_1 about the object plane is small compared to the distance of the object

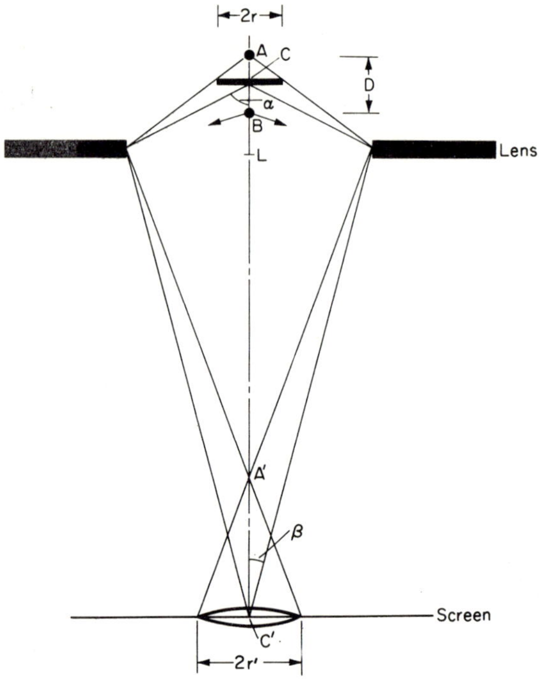

FIG. 69. Schematic diagram to illustrate the depth of field. (After G. Thomas, *Thin-Film Transmission Microscopy*, Wiley & Sons, fig. 55.)

from the lens as shown in Fig. 69. Since the angle subtended by rays leaving any small resolvable area of the specimen is α, D_1 is given by

$$D_1 = 2\delta/\alpha$$

Taking $\delta = 10$ Å and $\alpha = 5 \times 10^{-2}$ radians, D_1 is 1μ which is somewhat greater than that of the light microscope where, because α is very large, D_1 is about the same value as the resolution distance.

For a given lens the depth of focus D_2 is related to D_1 by the square of the magnification factor M, so that, in the final image, the depth of focus may be considered as infinite. In modern electron microscopes advantage is taken of this large depth of focus by recording the image on a photographic plate located some inches below the fluorescent screen, see Fig. 70, without impairing the focus.

Ciné techniques may also be used to record dynamic processes occurring while the specimen is under observation. In

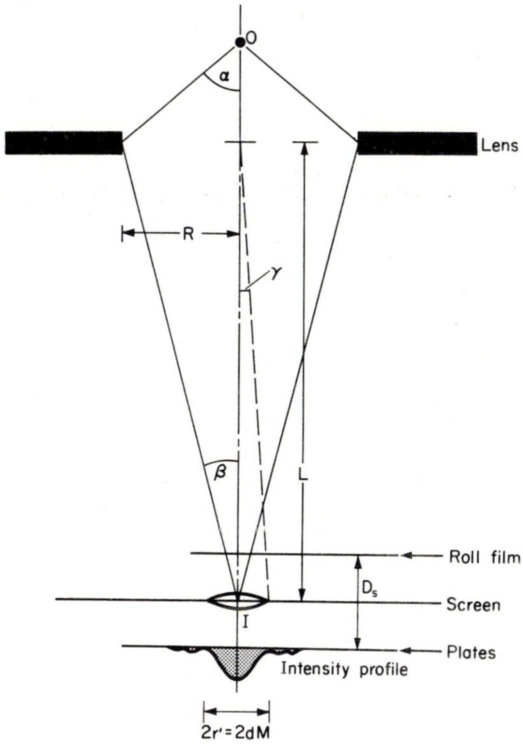

FIG. 70. Schematic diagram to illustrate the depth of focus. (After G. Thomas, *Thin-Film Transmission Microscopy*, Wiley & Sons, fig. 56.)

this case, in order to place the ciné camera near the viewing part of the microscope it is necessary to use a camera with a short focal length lens, e.g., *f*/0·95.

MICROSCOPE ATTACHMENTS

By far the most important operation in any modern microscope is tilting the specimen while it is under observation. Such a facility is necessary to bring the specimen into a strong diffracting position, since contrast from metals arises mainly from Bragg diffraction. Because of the short wavelength of electrons the important Bragg angles are all small ($\sim 1°$) so that extremely small tilts can alter the image contrast appreciably. The tilting device is clearly shown at X in Fig. 63(b).

For the correct interpretation of contrast effects it is often necessary to be able to tilt the specimen through large angles, say 30°, about more than one axis. It is possible to carry out such a tilting experiment by placing the specimen holder in a special goniometer stage which may be fitted into the microscope column. The ideal goniometer stage has two perpendicular tilt axes lying in the plane of the specimen and some, preferably automatic, means of recording both tilt angles whenever a photograph is taken. The important advantage of having the tilt axes co-planar with the specimen is that analyses of the results by stereographic projection is greatly facilitated. Unfortunately, few such stages are commercially available. There are, however, many stages which have one tilt axis and a 360° rotation; the E.M.6 goniometer stage is shown in Fig. 71. The general disadvantage of this type of stage which incorporates a 360° rotation is that the rotation axis does not remain perpendicular to the plane of the specimen. Consequently, in transferring the experimental data to a stereogram the tilt axis is not confined to the circumference.

Important information about a particular metal or alloy can often be obtained by carrying out certain thermal and mechanical experiments on the specimen while it is still under observation in the microscope. For example, by heating the grid

FIG. 71. Photograph of the goniometer stage for the E.M.6.

upon which the specimen sits with a small battery it is possible to study phase transformations, annealing behaviour, oxidation reactions, etc. By cooling the sample, high temperature metallurgical reactions can be prevented and low-temperature reactions induced, as well as contamination and oxidation of the specimen reduced. Cooling to about −120°C may be achieved merely by connecting one end of a copper rod to an insulated specimen holder while the other is immersed into a dewar of liquid nitrogen. More elaborate cooling, e.g., for superconductivity experiments, can be achieved by circulating a coolant around channels in the specimen holder.

The most difficult operation to carry out in the microscope is plastic deformation, particularly at different temperatures. Several straining devices have been used however, some based on the bimetallic strip principle and others on applying the load directly by moving a screwed rod.

Many electron microscopes may also be operated as reflection microscopes by inserting a reflection stage. Although not often used for metallurgical experiments, reflection microscopy is a useful technique for studying the surface topography of bulk specimens.

PREPARATION OF SPECIMENS

With the electron microscope aligned for operation in transmission, a metal specimen may be examined indirectly using replicas, or directly with thin sections or films (~1000 Å thick).

REPLICA TECHNIQUES

Thin replicas taken from the surface of the metal to display its contours and special topographical features are of considerable use in those fields of investigation where surface structure is important, e.g., fracture. Three different types of replicas are in general use, namely (i) plastic, (ii) carbon and (iii) oxide. With the plastic replica technique, a weak solution of a plastic in a volatile solvent (e.g., collodion in amyl acetate or formvar in chloroform) is dropped onto the surface of interest. The solvent evaporates and leaves a thin plastic film behind which can be stripped off using cellotape, to give a "negative" of the metal surface. In the carbon replica technique, carbon is evaporated onto the metal surface by resistance-heating carbon rods in an evacuated evaporating unit. The carbon film may be stripped by first scoring it into squares with a sharp knife and then etching or electro-polishing the specimen away beneath the film. The film is then picked-up out of the solution on a 100 mesh grid used to hold the specimen in the microscope.

A carbon film may also be used to extract precipitate particles from a multi-phase alloy. This technique, known as the carbon extraction technique, arises because etching or electro-polishing usually results in a dissolution of the matrix, with the result that the precipitate particles are left adhering to the carbon film. The particles extracted can then be examined by electron and X-ray diffraction techniques and identified. Figure 72(a) shows an example of carbon extraction replica.

In the oxide technique, used principally for aluminium and its alloys, an oxide film a few hundred Angstroms thick is formed on a clean electro-polished surface by anodizing in a suitable solution. The thin anodic film, which forms at a rate of ~14Å per volt, may be removed either by dissolving away the base metal or by electro-polishing after scoring the film into squares.

With these replicas, the contrast in the microscope image arises from variations in film thickness and dark contrast arises from the thicker areas. If extracted particles adhere to the replica then these will scatter the electrons by Bragg diffraction and additional dark contrast will also result. In aluminium alloys, second phases and precipitated particles oxidize at a different rate to that of the matrix metal, and this gives rise to variations in thickness of the anodic film and hence good contrast in the electron-optical image. Figure 72(b) shows an example of the Widmanstatten pattern produced by precipitated θ' particles in an age-hardened Al–4% Cu alloy.

THIN METAL FOIL TECHNIQUES

Since metals only become "transparent" to electrons when they are a few thousand Angstroms thick it is necessary to produce thin film sections either from bulk material or directly, for example by evaporation. Specimens cut from bulk material are clearly of greatest interest since it is then possible to study phenomena which have been produced in the bulk material, such as quenching, age-hardening, phase transformation, plastic deformation and irradiation damage. Nevertheless, specimens formed directly into thin sections, for example, pure metals by evaporation or oxide compounds by controlled oxidation, also give useful structural information.

In preparing microscope specimens (~1000 Å thick) from

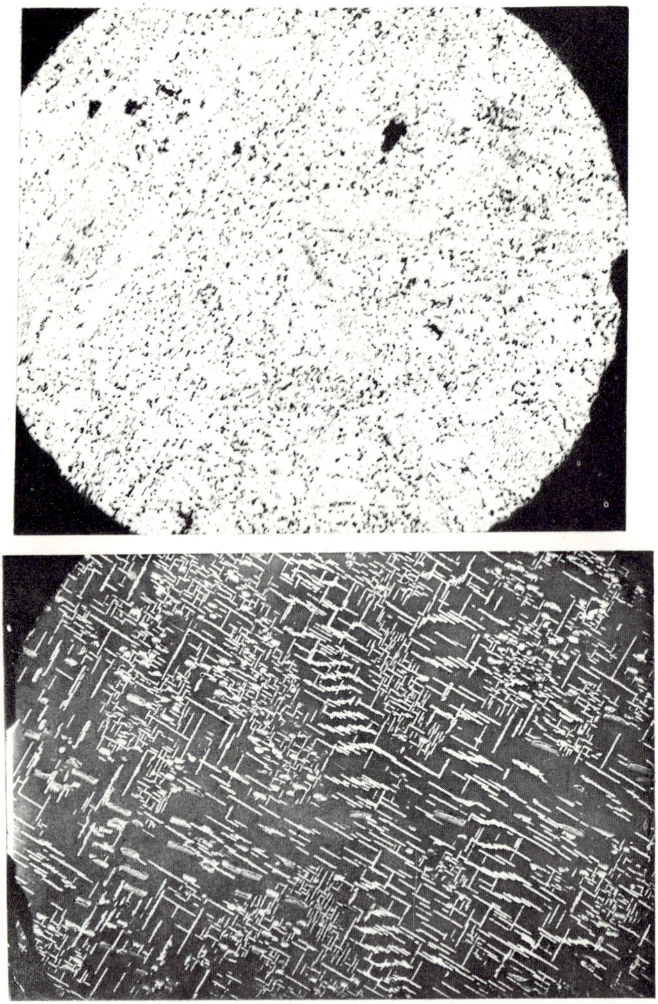

(a)

(b)

FIG. 72. (a) Carbon extraction replica of an alloy steel tempered at 650°C. (After D.A. Woodford, unpublished.) (b) Oxide-replica of θ'-precipitate in Al/4% copper alloy. (After G. Thomas and J. Nutting. *Proc. Int. Conf. on Electron Microscopy*, London, 1954, p. 222, fig. 2.)

bulk metals (e.g., tensile test pieces or single crystals) a thin slice is first removed by one of several techniques, namely (i) cutting with a diamond slitting wheel or acid saw, (ii) machining with a fine jet of a suitable electrolyte, (iii) local sand blasting using some abrasive, and (iv) spark erosion. With the spark-erosion technique a rapid series of spark discharges of controlled energy is produced between a tool and the specimen, and the spark erodes the specimen material away at a rate dependent on the energy and frequency of the discharges; high energies are used for fast roughing work and low energies for producing finer finishes. In one commercial instrument – Servomet – the correct gap between the specimen and the tool of the spark apparatus is controlled by a sensitive servo system. The technique is very useful for preparing samples suitable for final electro-thinning from hard brittle materials or for sectioning single crystals without introducing extensive mechanical damage.

The relatively thick section produced in this way is then finally thinned either by electro-polishing or chemical polishing in the appropriate solution. A typical arrangement for electro-polishing is shown schematically in Fig. 73(a), and Fig. 73(b) illustrates the conditions governing correct polishing. Preferential electro-polishing usually takes place at the edges of the sample and generally this is eliminated by coating them with a non-conducting lacquer such as "lacomit", which is later removed by dissolving in acetone. Electro-polishing is continued until holes appear in the "window" area and then the sample is removed from the electrolyte and washed in alcohol and distilled water. Specimens suitable for the microscope may then be cut from the thin edges around the holes with a sharp steel scalpel and mounted on 100 mesh copper grids so that the jagged edge of the foil lies across the centre of the grid (see Fig. 63a).

Thin metal specimens of the appropriate thickness may also be produced directly by evaporating the metal in the form of wire in vacuo and allowing the vapour to condense onto a cold substrate, such as mica or a cleaved crystal or rock-salt. Methods also exist for producing thin electro deposits from solution. It must be borne in mind, however, that the properties of thin foils produced by evaporation or deposition are usually different to those of the bulk metal.

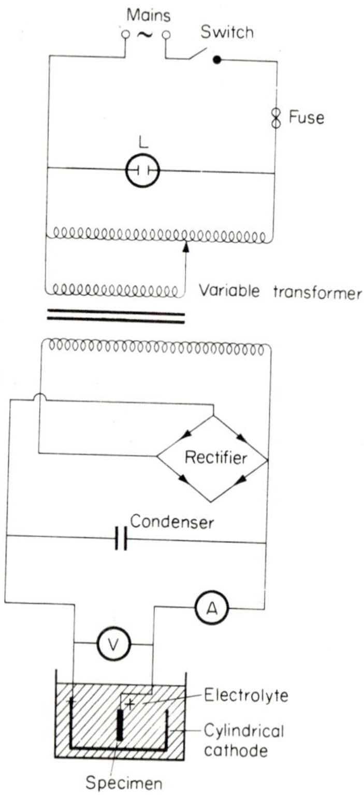

FIG. 73(a). Schematic diagram of electrical circuit.

For thin metal foils, the contrast seen in the electron optical image arises principally from Bragg scattering of the electron beam and not to any appreciable extent to differences in absorption due to variations in foil thickness. A concise account of the contrast mechanism in thin metal foils is given in Chapter VIII.

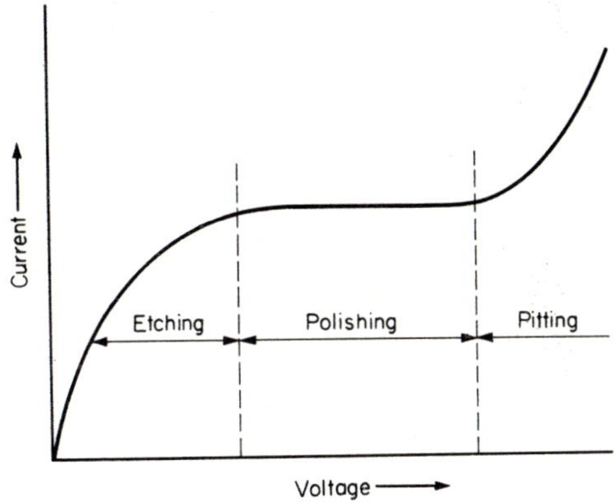

FIG. 73(b). The correct $i-V$ conditions for electro-polishing.

SUGGESTIONS FOR FURTHER READING

B. CHALMERS and A. G. QUARRELL, *The Physical Examination of Metals*, Arnold, 1960.

M. E. HARRIS and V. E. COSSLETT, *The Electron Microscope*, Spon Ltd., 1961.

G. THOMAS, *Transmission Electron Microscopy of Metals*, Wiley & Sons, 1962.

CHAPTER VIII

ELECTRON MICROSCOPY – II CONTRAST THEORY

INTRODUCTION

When an electron beam strikes a thin metal foil, there emerges from the other side two beams, one strongly diffracted* and the other strongly transmitted. Taking the intensity of the incident beam as unity and that of the diffracted beam as I, the intensity of the transmitted beam is $(1-I)$. If the incident beam passes through a crystal lattice near a defect, e.g., a dislocation, it encounters a localized change in diffraction conditions and the diffracted, and hence the transmitted, intensities are correspondingly modified. As a result, the defect can be distinguished from the background, i.e., it shows contrast. On a bright-field micrograph, regions, such as dislocations, which strongly diffract electrons appear dark on a bright background since the transmitted intensity from these regions is less than that from the surrounding perfect crystal. For a dark-field image, this contrast is reversed.

In order to explain the contrast observed at lattice defects, it is necessary to calculate the diffracted intensity from a perfect crystal and then try to see how this intensity is modified by the presence of lattice defects. These calculations permit the prediction of several important contrast effects and, since a knowledge of these contrast effects is necessary for an under-

*In addition to one strongly diffracted beam there are several weakly diffracted beams. Hence, the pattern of one strong and several weak spots obtained by S.A.D. (see Chapter IX).

standing of the defects which give rise to them, this chapter will be devoted to a simplified account of contrast theory.

DIFFRACTION OF ELECTRONS BY PERFECT CRYSTALS

To describe the diffraction contrast from lattice defects completely, a dynamical theory taking account of the equilibrium between all waves in the imperfect crystal is required. However a simpler, kinematical theory has been developed which, although it is an approximation justified only when the deviation from the reflecting position is large or when the crystal is very thin, does explain many contrast effects. We shall now outline the use of the kinematical thoery to predict the the intensity of the diffracted beam from a perfect crystal.

Let us represent an electron wave at the position **r** by the function exp $(2\pi i\mathbf{k}_0 \cdot \mathbf{r})$ where $\mathbf{k}_0$ is the electron wave vector of magnitude $1/\lambda$. When the electron wave is incident upon a crystal there will be an elastically scattered wave exp $(2\pi i\mathbf{k} \cdot \mathbf{r})$, the amplitude of which is obtained by adding up the scattered amplitudes from all the atoms in the crystal. Thus

$$A = f(2\theta) \sum_n \exp\left[-2\pi i(\mathbf{k}_0 - \mathbf{k}) \cdot \mathbf{r}_n\right] \tag{8.1}$$

where $f(2\theta)$ is the electron-scattering factor for the appropriate angle and the summation extends over all atomic positions in the crystal. The terms in the summation simply take account of the different path lengths for rays scattered by different atoms. As shown previously in Fig. 50 the path difference between two scattered rays in the direction **k** is $A_1 N - A_2 M$, the difference between the projections of $A_1 A_2$ on the directions of scattering and incidence. In vector notation this path difference is the difference between the scalar products of **r** and **k** and **r** and $\mathbf{k}_0$, or

$$\begin{aligned} \text{path difference} &= \mathbf{r} \cdot \mathbf{k} - \mathbf{r} \cdot \mathbf{k}_0 = \mathbf{r} \cdot (\mathbf{k} - \mathbf{k}_0) \\ &= \mathbf{r} \cdot \mathbf{K} \end{aligned} \tag{8.2}$$

Thus the amplitude of electron waves scattered from an assembly of unit cells is

$$A = \sum_{n} F_n \exp[-2\pi i \mathbf{K} \cdot \mathbf{r}_n] \tag{8.3}$$

where $\mathbf{r}_n = n_1\mathbf{a} + n_2\mathbf{b} + n_3\mathbf{c}$. The total scattered amplitude is greatest when all the terms in the summation have the same phase (i.e., when $\exp[-2\pi i\ \mathbf{K} \cdot \mathbf{r}_n] = 1$ for all values of $\mathbf{r}_n$) which is the condition when the Bragg law is satisfied and

$$\mathbf{K} = h\mathbf{a}^* + k\mathbf{b}^* + l\mathbf{c}^* = \mathbf{g} \tag{8.4}$$

as shown in Fig. 74(a). The vector **g** is normal to a crystal plane with spacing $d = 1/\mathbf{g}$ and hence for strong diffraction $d = 1/|K| = \lambda/2 \sin\theta$, which is the Bragg law.

The electrons are, of course, scattered over a range of angles and this can be discussed in terms of a distribution of intensity in reciprocal space. In general, the crystal will not be at the exact Bragg reflecting position and then, as shown in Fig. 74(b),

$$\mathbf{K} = \mathbf{g} + \mathbf{s} \tag{8.5}$$

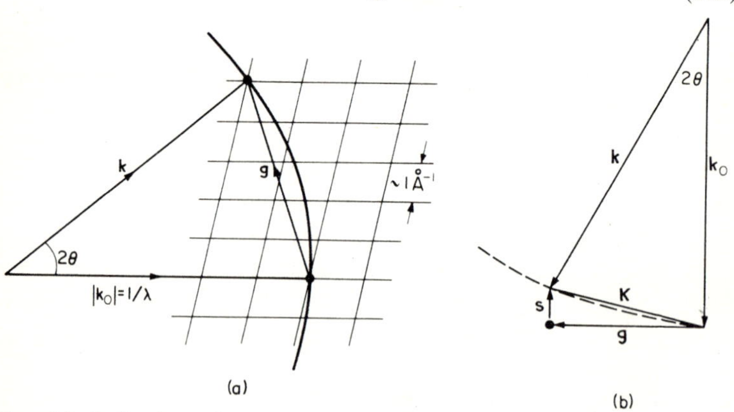

FIG. 74. Reflecting sphere construction for diffraction when the reciprocal lattice point is (a) on and (b) off the reflecting sphere. With the reciprocal lattice point outside the reflecting sphere, as shown in (b), s is negative.

where **s** is a small vector which describes the deviation from the Bragg angle. In this case,

$$A = \sum_n F_n \exp\left[-2\pi i(\mathbf{g}+\mathbf{s}) \cdot \mathbf{r}_n\right] \tag{8.6}$$

and for a **perfect crystal** this equation reduces to

$$A(s) = \sum_n F_n \exp\left[-2\pi i\mathbf{s} \cdot \mathbf{r}_n\right] \tag{8.7}$$

because **g** and $\mathbf{r}_n$ are lattice vectors and hence $\mathbf{g} \cdot \mathbf{r}_n$ is an integer. This equation is usually written as an integral of the form

$$A(s) = \frac{F}{V}\int \exp\left[-2\pi i\mathbf{s} \cdot \mathbf{r}_n\right] d\tau \tag{8.8}$$

by assuming all cells are alike, where V is the volume of a unit cell and dτ is a volume element. For a rectangular parallelapiped crystal of sides A, B and C

$$A(s) = \frac{F}{V} \cdot \frac{\sin \pi Au}{u} \cdot \frac{\sin \pi Bv}{v} \cdot \frac{\sin \pi Cw}{w} \tag{8.9}$$

where u, v and w are the components of **s** along the directions of A, B and C.

Each of the sine terms in equation (8.9) is of the type met with in the theory of an ordinary plane diffraction grating, exhibiting both main and subsidiary maxima. The intensity variation along u, for v and w equal to zero, has a main half-peak width $u = 1/A$ and thus for a spherical crystal radius R the distribution of appreciable intensity is spherically symmetrical about a reciprocal lattice point and has a half-peak width of $1/R$.

In electron microscopy the specimen is usually a thin platelet normal to the electron beam, and consequently the intensity distribution is spread out in a spike normal to the crystal plate with a width $1/t$, where t is the crystal thickness. This is shown schematically in Fig. 75. If $t = 200$ Å and $d \simeq 2$ Å the range of reflection is $2\Delta\theta = 2(\frac{1}{2}t)/(1/d) = d/t \simeq 10^{-2}$ radians. Thus the angular range over which the crystal may be misoriented and yet still diffract is of the order of the Bragg reflection. Consequently it is possible to get simultaneous reflections operating at one instance; this is quite different to the behaviour of a crystal diffracting X-rays. With increase in thickness of the foil t, θ decreases but there is a limit to the decrease because the beam is completely diffracted away after only a few hundred Angstroms of thickness, so that the number of crystal planes contributing to the reflection is strictly limited. For such thick crystals the range of reflection is given by $\theta = d/t_0$, where t_0 is known as the extinction distance* and is of the order of 200–600 Å for 100 kV electrons reflected from low order planes in metals (see Table 4).

* t_0, the extinction distance for electrons is given by the expression

$$t_0 = \frac{\pi V \cos\theta}{\lambda F_e(\theta)} \tag{8.11}$$

where V is the volume of the unit cell, λ is the wavelength of the electrons and $F_e(\theta)$ is the structure factor for electrons. $F_e(\theta)$ is given by

$$F_e(\theta) = \Sigma f_e(\theta) \exp 2\pi i(hu_i + kv_i + 1w_i)$$

where $f_e(\theta)$ is the atomic scattering factor for electrons.

$$f_e(\theta) = 0.023934\,\{Z - f_x(\theta)\}\left(\frac{\sin\theta}{\lambda}\right)^2$$

where Z is the atomic number of the metal concerned and $f_x(\theta)$ is the atomic scattering factor for X-rays. $f_x(\theta)$ is listed as a function of $\frac{\sin\theta}{\lambda}$ on p. 218 of *International Tables for X-ray Crystallography*, vol. III. $\frac{\sin\theta}{\lambda}$ for the reflection corresponding to that for which t_0 is being calculated is derived from the Bragg equation; $\frac{\sin\theta}{\lambda} = \frac{1}{2d}$ for first order reflections.

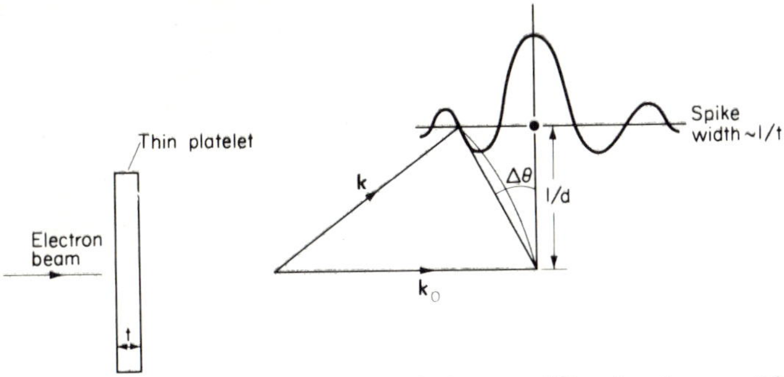

FIG. 75. Schematic representation of electron diffraction from a thin crystal platelet normal to the electron beam.

TABLE 4. EXTINCTION DISTANCES (Å) FOR LOW ORDER REFLECTIONS WITH 100 kV ELECTRONS

Reflection	Al	Ni	Mg	Zn	V	Fe	Mo	W
110	–	–	–	–	380	295	260	180
111	640	250	–	–	–	–	–	–
200	740	300	–	–	545	445	385	250
211	–	–	–	–	700	590	465	310
220	1240	470	–	–	–	–	–	–
0002	–	–	935	380	–	–	–	–
$01\bar{1}1$	–	–	1155	500	–	–	–	–
$01\bar{1}0$	–	–	1770	785	–	–	–	–

Multiplication of the expression for the amplitude by its complex conjugate yields a quantity proportional to the diffracted intensity, *I*. Most of the intensity is concentrated about the reciprocal lattice point and thus we may assume that at the lower surface of a platelet specimen, such as that shown in Fig. 76, the diffracted and transmitted beams are planar in the directions of diffraction and incidence respectively. Under these conditions the intensity diffracted may be approximated

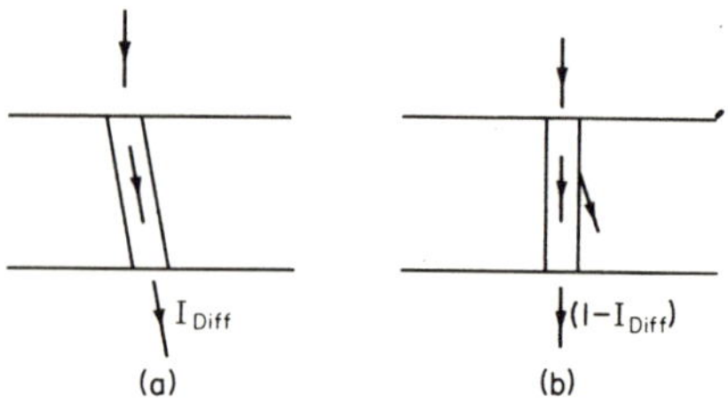

FIG. 76. Column approximation for (a) dark-field and (b) bright-field calculation.

to that diffracted from a column in the direction of the diffracted beam, as shown in Fig. 76(a); for the transmitted beam the intensity is $1-I$ as shown in Fig. 76(b). The diffracted intensity from such a column is

$$I \propto \frac{\sin^2(\pi t s)}{(\pi s)^2} \tag{8.10}$$

and from this equation it will be noted that the intensity of the transmitted and diffracted beams oscillate with depth in the crystal as illustrated in Fig. 77(a). At successive depths t_0 through the crystal, the diffracted intensity is zero, i.e., extinction occurs. The depth periodicity of the intensity is $1/s$ and the maximum wavelength of the oscillations is the extinction distance.

The theory so far discussed for perfect crystals is able to account for two contrast features, namely thickness contours and extinction contours commonly observed in transmission electron microscopy, neither of which arises from the presence of lattice defects. A common example of thickness contour contrast is the observation of fringes in wedge-shaped crystals (see Fig. 77b), which arise from the sinusoidal variation with thickness of the transmitted beam discussed above. **s**, the deviation from the Bragg reflecting condition, is constant

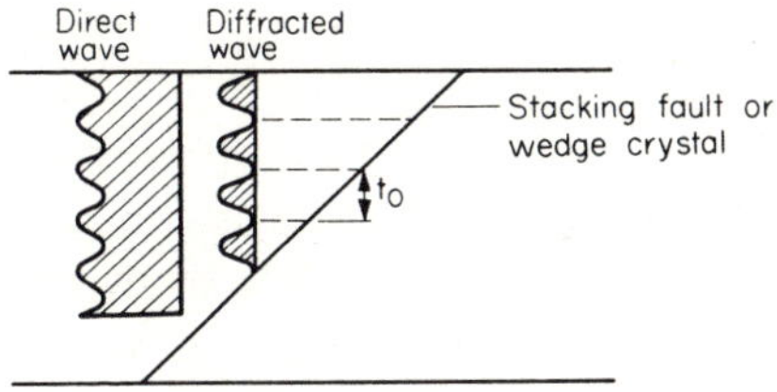

(a)

(b)

FIG. 77. (a) Schematic diagram showing the oscillation of transmitted and diffracted intensity in a thin-metal foil. The distance t_0 is the extinction distance. (b) Fringes from a wedge-shaped crystal.

throughout the wedge. Bend or extinction contours arise in buckled foils, since there is always a narrow region of crystal, extending across the buckled area, which is favourably oriented for Bragg reflection (see Fig. 90). This region shows up on a micrograph as a contour which, in the case of very thin foils, has subsidiary contours associated with it. The presence of these subsidiary bend contours is due to the variation of **s** across the buckled area, and can be represented by an intensity profile as shown in Fig. 78(a). The dark-field image is symmetric about $\mathbf{s} = 0$, but in bright-field, the image is asymmetric about $\mathbf{s} = 0$ with a so-called **anomalous transmission** region for a certain small and positive value of **s**. The contrast at lattice defects is best in this anomalous transmission region. In thick foils, the subsidiary intensity maxima and minima are reduced by absorption effects giving the profiles shown in Fig. 78(b). As a result, only the main bend contour is seen.

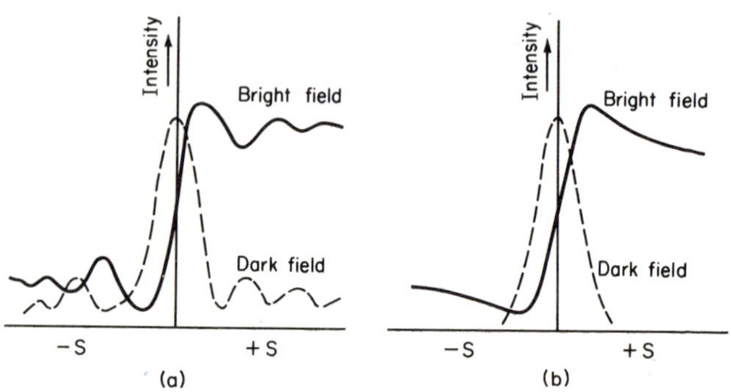

FIG. 78. Transmitted intensity profiles for (a) thin and (b) thick foils.

ELECTRON DIFFRACTION BY AN IMPERFECT CRYSTAL

To understand how the contrast arises from lattice defects, it is necessary to determine how the expression for the intensity of the beam diffracted from a perfect crystal is altered by the presence of lattice defects.

When the crystal is **imperfect** certain atoms around the lattice defects are displaced from their ideal atomic positions. If the atom at $\mathbf{r}_n$ is displaced by a vector $\mathbf{R}$, the amplitude of the wave diffracted by this atom is multiplied by an additional phase factor $\exp[2\pi i\mathbf{K}\,.\,\mathbf{R}]$, so that from equation (8.7), based on the kinematical theory, the amplitude scattered by an imperfect crystal is

$$A(s) = \Sigma F_n \exp\left[-2\pi i(\mathbf{g}+\mathbf{s})\,.\,(\mathbf{r}_n+\mathbf{R}_n)\right] \qquad (8.12)$$

$$= \Sigma F_n \exp\left[-2\pi i\mathbf{s}\,.\,\mathbf{r}_n\right] \exp\left[-2\pi i\mathbf{g}\,.\,\mathbf{R}_n\right] \qquad (8.13)$$

where $\mathbf{s}\,.\,\mathbf{R}$ has been neglected because it is small, and also $\mathbf{g}\,.\,\mathbf{r}_n$ because it is an integer and therefore does not affect the phase. Equation (8.13) may also be expressed as an integral,

$$A(s) = \frac{F}{V}\int_{\text{crystal}} \exp\left[-2\pi i\mathbf{s}\,.\,\mathbf{r}_n\right] \exp\left[-2\pi i\mathbf{g}\,.\,\mathbf{R}_n\right] d\tau \qquad (8.14)$$

and comparing this with equation (8.8) it is clear that the image contrast from lattice imperfections, e.g., dislocations, stacking faults, etc., is essentially a phase contrast effect in the Bragg reflection, the phase factor $\alpha = 2\pi\mathbf{g}\,.\,\mathbf{R}_n$ arising from the atomic displacements $\mathbf{R}$. The total amplitude diffracted by a column of imperfect crystal, as before, is proportional to

$$\int_0^t \exp\left[-2\pi i(sz+\mathbf{g}\,.\,\mathbf{R})\right] dz \qquad (8.15)$$

where the integration is over a co-ordinate z in a column of length t (see Fig. 79) and the quantity $\mathbf{s}\,.\,\mathbf{r}_n$ in the phase factor reduces to sz (with s equal to the z-component of s) because the lateral dimensions of the column are not considered. Multiplication of equation (8.15) by its complex conjugate yields a quantity proportional to the intensity I, diffracted from a region of crystal containing a defect such as a dis location. The difference between this intensity and that diffracted by the surrounding perfect crystal, gives rise to the contrast observed at the defect on a dark field image. According to the kinematical theory, the bright-field image is complimentary to the dark-field image since the transmitted intensity is $(1-I)$.

In practice, this prediction that dark and bright-field images are complimentary is found not always to be true and illustrates one of the major limitations of the kinematical theory. The reason for this is that the kinematical treatment completely neglects absorption, and the fact that the diffracted wave in a crystal at the Bragg setting is also incident on the reflecting planes at the Bragg angle, so that multiple reflection can occur. The dynamical theory takes these factors into consideration. Another contrast effect not predicted by the kinematical theory is found in thick foils, where dislocations inclined to the surface exhibit a black and white zig-zag type of contrast which reverses at different depths in the specimen. The effect of

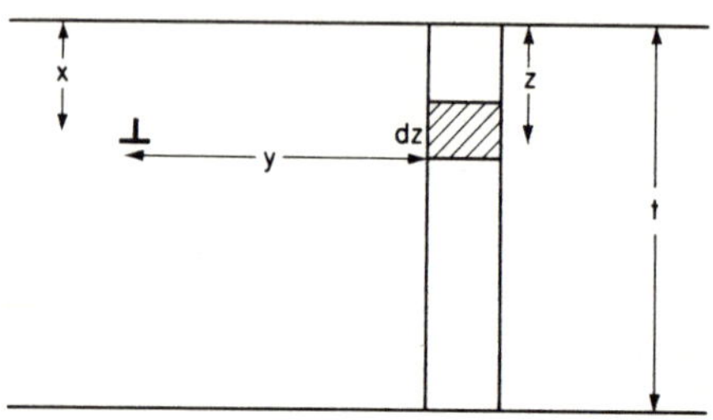

FIG. 79. Column approximation for a crystal containing a dislocation.

absorption is to cause this zig-zag contrast to die away in the centre of the foil. (The zig-zag contrast seen on dislocations near the surface of the foil should not be confused with any real structural effect such as dislocation pinning.) A similar effect is observed from stacking fault contrast in foils of thickness $t > 6t_0$. A fault on a plane inclined to the foil surface gives rise to a fringe contrast (see next chapter) and the effect of absorption, not predicted by kinematical theory, is to destroy the fringe contrast in the middle of the fault, and to cause the breakdown of the complementary nature of the bright- and dark-field images. It is found that the fringe images are still complementary for that part of the fault near the bottom of the foil but for the upper surface of the foil the images are similar and not complementary. This effect is further discussed in Chapter IX.

The kinematical predictions do hold for conditions not too close to Bragg reflection since here the dynamical equations reduce to those given by the kinematical theory.

SUGGESTIONS FOR FURTHER READING

HIRSCH, P. B., HOWIE, A., and WHELAN M. J., *Phil. Trans.* A, **252,** 499 (1960).

A. HOWIE, Quantitative experimental study of dislocations and stacking faults by transmission electron microscopy, *Metallurgical Reviews*, **6,** 467 (1961).

P. B. HIRSCH, A. HOWIE and M. J. WHELAN, A kinematical theory of diffraction contrast of electron transmission microscope images of dislocations and other defects. *Phil. Trans. Royal Soc.* A, **252,** 499 (1960).

H. HASHIMOTO, A. HOWIE and M. J. WHELAN, Anomalous electron absorption effects in metal foils, *Phil. Mag.*, **5,** 967 (1960).

A. HOWIE and M. J. WHELAN, Diffraction contrast of electron microscope images of crystal lattice defects: II. The development of a dynamical theory, *Proc. Roy. Soc.* A, **263,** 217 (1961).

A. HOWIE and M. J. WHELAN, Diffraction contrast of electron microscope images of crystal lattice defects: III. Results and experimental confirmation of the dynamical theory of dislocation image contrast, *Proc. Roy. Soc.* A, **267**, 206 (1962).

CHAPTER IX

ELECTRON MICROSCOPY – III INTERPRETATION

INTRODUCTION

It was explained in Chapter VIII that image contrast from a crystal imperfection arises from the additional phase factor $\alpha = 2\pi \mathbf{g} \cdot \mathbf{R}$ introduced by the defect into the equation for the intensity of the diffracted beam. This chapter will be devoted to the various ways in which a study of this phase factor can be used to determine the nature of lattice defects. However, all the information that is necessary to deduce the phase factor is contained in the selected area diffraction pattern taken from the area containing the defect concerned, and consequently the first section will be devoted to the interpretation of S.A.D. patterns.

ANALYSIS OF SELECTED AREA DIFFRACTION PATTERNS

The usual type of transmission electron diffraction pattern from a single crystal region is a cross-grating pattern of the form shown in Fig. 68. The simple explanation of the pattern can be given by considering the reciprocal lattice and reflecting sphere construction previously discussed in Chapter V. In electron diffraction the electron wavelength is extremely short ($\lambda = 0{\cdot}037$Å at 100 KV) so that the radius of the Ewald reflecting sphere is about 25Å^{-1}, which is about 50 times greater than **g**, the reciprocal lattice vector. Moreover, because λ is small the Bragg angles are also small (about 10^{-2} radians or $\sim\frac{1}{2}^{\circ}$ for low order reflections) and hence the reflection sphere may be considered as almost plane in this vicinity.

If the electron beam is closely parallel to a prominent zone axis of the crystal then several reciprocal lattice points (somewhat extended because of the limited thickness of the foil) will intersect the reflecting sphere, and a projection of the prominent zone in the reciprocal lattice is obtained, i.e., the S.A.D. pattern is really a photograph of a reciprocal lattice section. Figure 80(a) shows the prominent zone which is projected

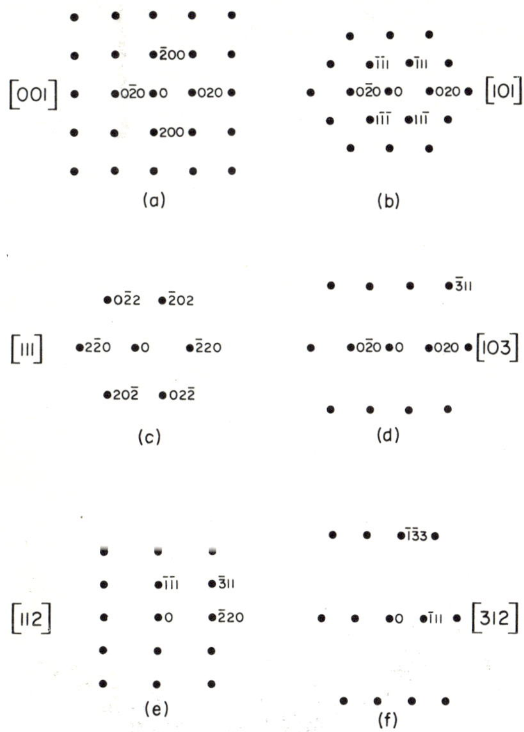

FIG. 80. Prominent cross-grating patterns observed from a f.c.c. lattice. By inspection of the pattern it is possible to determine an approximate direction of incidence of the electron beam.

when the electron beam is parallel to [001] in a f.c.c. lattice. Only one spot is usually strong,* however, and this corresponds to the strongly diffracted beam diffracted by the crystal planes which are exactly, or almost exactly, oriented for Bragg reflection (see Figs. 81 and 94). To ensure that only one strong reflection is operating in a bright-field image, it is good practice to take S.A.D. patterns only from areas straddled by bend or extinction contours; each bend contour maps out a region where the foil is so buckled or bent that the electrons passing through it are strongly scattered into Bragg reflection. An example is shown in Fig. 81. The various extinction contours visible in this micrograph give rise to different Bragg reflections in the cross-grating pattern.

The size of electron diffraction spots on S.A.D. patterns of strain-free regions is governed almost entirely by the beam divergence $d\theta$. It is therefore possible to obtain very high resolution on the S.A.D. patterns by reducing $d\theta$ sufficiently by defocusing the condenser lens. However, if the foil is not strain-free, the spots may be broadened by the strains. This effect is noticed in patterns taken from regions of very high dislocation density.

There are four ways of analysing S.A.D. patterns, all of which should be used to cross-check each other. Each of these methods will now be explained in detail.

1. *Inspection method.* A S.A.D. pattern can usually be identified by comparing it with known reciprocal lattice sections of the structure being studied. Reciprocal lattice models or diagrams of common reciprocal lattice sections (e.g., Fig. 80) are useful aids. It often happens, however, that

*It sometimes happens that more than one fairly strong spot appears on the S.A.D. pattern and while one of the spots is usually more intense than the others, it is essential in this case to examine the dark-field image obtained by using that particular strongly diffracted beam. A contrast reversal will be obtained if the diffracted beam chosen is responsible for imaging the defect under examination.

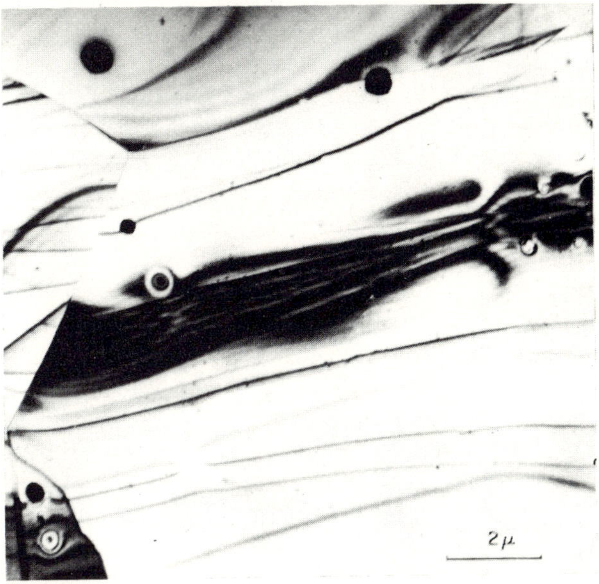

(a)

(b)

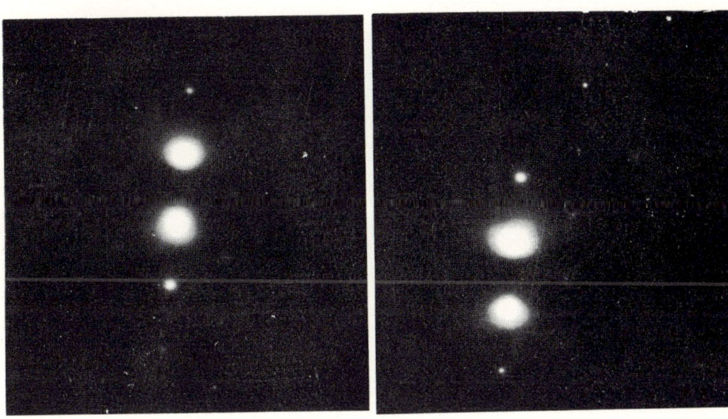

(c)

FIG. 81. A bend contour in a thin foil of vanadium (a) and selected area diffraction patterns taken from either side of it (b) and (c). (J. W. Edington and R. E. Smallman, unpublished.)

a well defined cross-grating pattern is not obtained in which case little can be learned by inspection.

2. *Camera constant method.* A ray diagram showing how a diffraction spot is obtained on a photographic plate is shown schematically in Fig. 82. Since θ is small and approximately equal to $\sin\theta$, the Bragg equation $\lambda = 2d \sin\theta$ can be written as

$$\lambda = d \,.\, 2\theta$$

where d is the d-spacing corresponding to the spot being indexed. Substituting $2\theta = \tan 2\theta = r/L$, we get the equation

$$L\lambda = \mathrm{d}r \qquad (9.1)$$

where L is the so-called camera length and r is the distance from the centre of the S.A.D. pattern, as defined by the spot

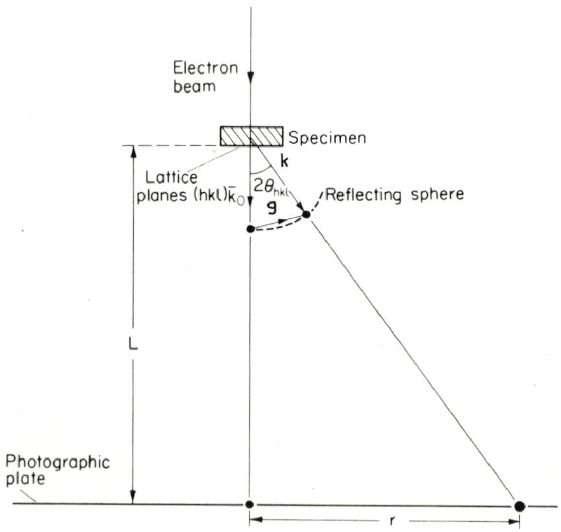

FIG. 82. Schematic diagram showing the relation between the camera length L and the diffracted spot on the S.A.D. pattern.

corresponding to the transmitted beam, and the spot to be indexed. The term $L\lambda$ is known as the camera constant and is determined from measurements of r on the S.A.D. pattern of a substance of accurately known lattice parameter, e.g., the ring pattern of polycrystalline gold. The units of $L\lambda$ are mm Å if r is measured in mm, d having the units of Å. The procedure to index an S.A.D. pattern using the camera constant method is to measure the r value for each spot, calculate its d-spacing from equation (9.1) and compare this with the d-spacings listed on the ASTM card for the metal being studied. Only an approximate agreement can be expected since the camera length, and hence the camera constant, can vary considerably depending on the exact position of the specimen and the power of the lenses. The indices of the plane whose d-spacing best matches the observed one are then tentatively assigned to the spot concerned.

3. *Method of ratios.* These tentative indices should now be checked by comparing the measured values for r with the known d-spacings. According to equation (9.1), $r \propto (1/d)$. Therefore, the ratio of the r-values for any two spots

$$\frac{r_1}{r_2} = \frac{d_2}{d_1}.$$

If this ratio holds for every pair of spots, then the tentative indices must fit as far as the d-spacings are concerned. So far, however, no check has been made of the angles between the planes producing the diffraction spots. Such a check is essential and can best be done by stereographic projection.

4. *Stereographic projection.* The reciprocal lattice section to which a S.A.D. pattern corresponds will be evident when all the spots have been indexed as above. The reciprocal lattice section normal is then plotted on a standard stereographic projection for the crystal structure concerned as shown

in Fig. 88. The trace of the plane given by this normal should then be drawn and inspected to make sure that the poles of all the planes indexed lie upon it. The angle between any two of these poles is the angle between the planes they represent and should equal the angle subtended at the centre of the S.A.D. pattern by the corresponding spots. If this is found to be so, then the tentative indices are definitely established as correct.

Information from the indexed S.A.D. pattern can now be transferred to the micrograph. The indices of each spot define the trace, in the plane of the micrograph, of the plane normal to the vector drawn from the centre of the pattern to that spot. Consequently, when the S.A.D. pattern has been correctly oriented with respect to the micrograph (see Chapter VII for the rotation between the two), these traces can be drawn on the micrograph. The planes these traces represent are, of course, perpendicular to the plane of the micrograph. Thus the complete orientation of the micrograph has been determined.

The other important factor to be transferred from the S.A.D. pattern to the micrograph is the direction and sense of **g**, the reciprocal lattice vector of the set of lattice planes producing the contrast. The positive direction of this vector is from the centre of the S.A.D. pattern to the strong diffraction spot, and can be transferred to the micrograph when the latter is oriented correctly with its pattern. The indices of **g** have been previously determined as described above.

INTERPRETATION OF ADDITIONAL FEATURES ON S.A.D. PATTERNS

Apart from the pattern of spots, S.A.D. often produces additional features in the form of extra spots and Kikuchi lines. These features often provide information which is of vital importance to the interpretation of micrographs. In this section, we shall discuss the origin of these features and the information which can be obtained from them.

A S.A.D. pattern containing extra spots should always be carefully examined to see whether any of the spots arise from forbidden reflections. In electron diffraction, reflections which are forbidden by X-ray structure factor calculations may be observed by double diffraction. For example, the 222 spot in germanium is forbidden and yet is commonly observed by the diffraction of a 331 reflection in the $(11\bar{1})$ plane, thus

$$(222) = (331) - (11\bar{1}).$$

Apparently forbidden reflections also arise from spiked reciprocal lattice points in the section of reciprocal lattice above the one being indexed. This spiking of reciprocal lattice points is due to a relaxation of a Laue condition originating from a small crystal dimension in real space, i.e., a thin platelet.

Extra spots having a similar origin to those from forbidden reflections are formed when the specimen consists of two different crystals, e.g., a metal and its surface oxide. In this case, we get diffraction of the beam in one crystal by planes in the other, as shown in Fig. 83(a). As a result, extra spots are

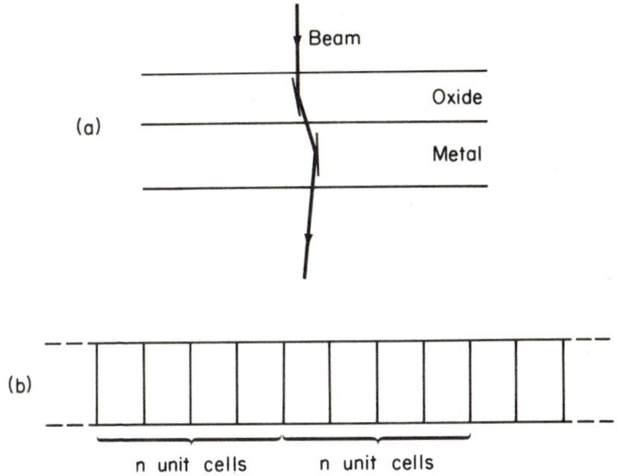

FIG. 83.

produced which correspond to $1/d_1 \pm 1/d_2$ where d_1 and d_2 are the d-spacings for the two crystals. The transmission electron micrograph due to double diffraction from two crystals is characterized by Moiré fringes, and the Moiré spacing is given by the relation

$$D = \frac{d_1 d_2}{\sqrt{d_1^2\alpha^2 + (d_2 - d_1)^2}} \tag{9.2}$$

where α is the angle of rotation, about the electron beam, between the two crystals. If $d_1 = d_2$, i.e., the two crystals are of the same substance, $D = \frac{d}{\alpha}$. If the crystals are different but there is no rotation (as often happens if an epitaxial relationship exists between the crystals), $D = d_1 d_2/(d_2 - d_1)$. An important property of Moiré fringes is that the spacing, given in equation (9.2), is independent of λ, the wavelength of the electrons, and hence a test for Moiré fringes is to switch from 100 kV to 80 kV when the spacing should not alter.

Superlattice reflections also lead to the formation of extra spots on S.A.D. patterns. Consider a superlattice which consists of n unit cells as shown in Fig. 83(b). In addition to the normal spots with spacings corresponding to $1/d$, weak superlattice spots are observed which can be indexed with spacings corresponding to $1/nd$. The brightness of superlattice spots is determined by the structure factor.

When the S.A.D. aperture straddles a sub-grain boundary or a twin boundary, spots are obtained from the crystal on both sides of the boundary, i.e., two sets of spots are produced. Such a pattern can yield information about the nature of the boundary, e.g., the angular misorientation across it and the rotation in the plane of the boundary. In the case of twin boundaries, this information permits an indentification of the twinning elements of the crystal. The occurrence of these doublet diffraction spots is limited in practice to cases where the misorientation is greater than about $\frac{1}{2}°$ for the following reason. When the condenser is defocused the beam divergence $d\theta$ is about $4{\cdot}3 \times 10^{-4}$ and, since the effective camera length of the instrument is about 50 cm, the size of the diffraction spots is about 0·2 mm. An order of magnitude of the distance of a double spot from the centre of the pattern is about 20 mm so that the least resolvable rotation is about $\frac{0 \cdot 2}{20} \times \frac{180}{\pi} =$ 0·6°, or approximately $\frac{1}{2}$°. The resolution is slightly better for higher order reflections since these are more distant from the central spot. An important requirement for the above analysis is that the misorientation is such that similar reflections occur on either side of the subgrain or twin boundary. For small misorientation angles this condition is usually satisfied.

S.A.D. patterns taken from moderately thick areas, $t \sim 4t_o$ to $6t_o$, which are relatively free from strains due to buckling usually contain Kikuchi lines. These take the form of striations, typical examples of which are shown in Fig. 84. The absence

FIG. 84. Kikuchi lines on a selected area diffraction pattern.

of buckling is indicated on the micrograph by the absence of extinction contours. Figure 85 illustrates the formation of these Kikuchi lines. The incident electron beam is assumed to be scattered incoherently throughout the volume of the

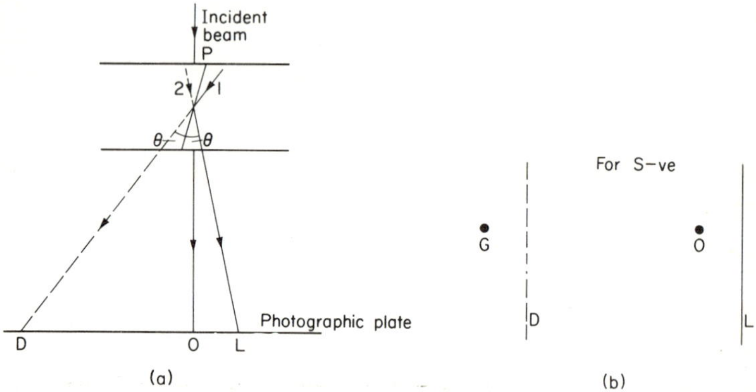

FIG. 85. Illustration of (a) the formation of Kikuchi lines and (b) the relation between Kikuchi lines L, D and the diffraction spots O and G.

crystal. The divergent incoherent waves give rise to the background on the electron diffraction pattern, and in addition are capable of being diffracted by crystal lattice planes. Thus, the incoherent waves which travel in the direction 1 in the crystal, i.e., at the Bragg angle θ to the crystal plane P, are scattered away from the point D on the photographic plate to the point L. The locus of L on the photographic plate is therefore the intersection of a cone of semi-vertical angle $\pi/2-\theta$ with the photographic plate. In practice θ is small, so that the intersection appears as a straight line. Electron waves propagating in direction 2, also at the Bragg angle θ to the plane P, are scattered away from L to D. However, because the distribution of the incoherently scattered waves is peaked in the forward direction (i.e., there are more electrons 2 than electrons 1) the point D receives an excess of electrons and L a deficit.

Thus a dark line appears on the photographic plate at *D* and a light line at *L* closest to the central spot.

Each pair of Kikuchi lines, *D* and *L*, corresponds to a Bragg reflection for the inelastically scattered electrons. They run in a direction normal to the **g** vector associated with their diffraction spot and are separated by a distance equal to $|\mathbf{g}|$ as shown in Fig. 85(b).

At the exact reflecting condition, i.e., when the reciprocal lattice spot lies on the reflecting sphere and therefore $\mathbf{s} = 0$, the Kikuchi lines pass through the central transmitted spot of the S.A.D. pattern and the operating diffraction spot respectively. This and the appearance of the Kikuchi lines for positive and negative **s** are summarized in Fig. 86.

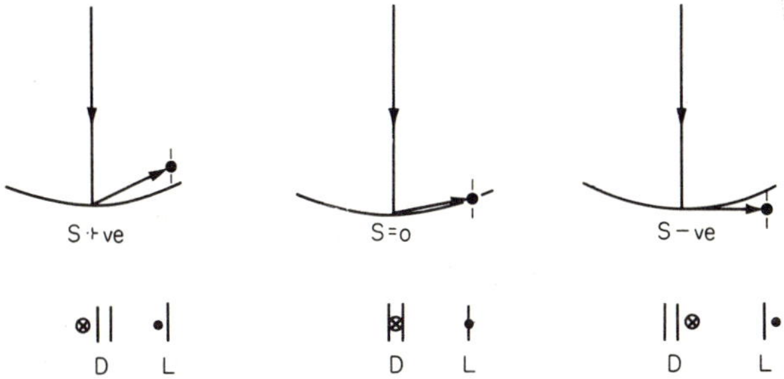

FIG. 86. Diagram showing the relation between the sign of **s** and the relative positions of dark and light Kikuchi lines.

Referring to Fig. 85, it is evident that the line of intersection of the crystal plane *P* with the photographic plate at *P*′ is then half-way between the Kikuchi lines *D* and *L*. Moreover, since the cones of diffracted waves at *D* and *L* are rigidly fixed to the specimen, relatively large movements of the Kikuchi lines on the photographic plate occur for small changes in orientation. This is in contrast to the behaviour of the S.A.D. pattern which does not change appreciably for small changes in orientation. Observations of Kikuchi lines can therefore give some indication of the accuracy of an orientation determined from the S.A.D. pattern. The deviation $\Delta\theta$ from the Bragg angle θ can be obtained from the relationship

$$\Delta\,\theta = 2f\theta$$

where f is the fractional distance of the dark line from the central spot (f is a fraction of **g**).

An accurate method for measuring the misorientation across a low angle boundary uses Kikuchi lines and gives the component misorientation angle in a plane perpendicular to the incident electron beam. Kikuchi lines formed in sub-grain boundary regions are split into doublets, the components of which arise from opposite sides of the subgrain boundary. Since the Kikuchi line pattern is fixed with regard to the specimen (see above), small changes in orientation across the boundary produce relatively large separations of the components of the doublets. The separation of each line gives the resolved component of the small misorientation angle in the direction of the Kikuchi line. The component perpendicular to this direction does not produce any noticeable effect on the line. Misorientation of less than $\frac{1}{2}°$ can be detected by this method.

INDEXING CRYSTALLOGRAPHIC FEATURES ON MICROGRAPHS

A slip plane can often be identified by contrast sometimes seen along the lines of intersection of the slip plane with the top and bottom surfaces of the foil, the contrast over the rest of the slip plane remaining the same as for the background. An example is shown in Fig. 87(a). Other crystallographic features which produce line contrast where they intersect the foil surfaces are cracks and twin boundaries. The planes on which such features lie can usually be indexed by transferring the intersections to a standard stereographic projection. Thus,

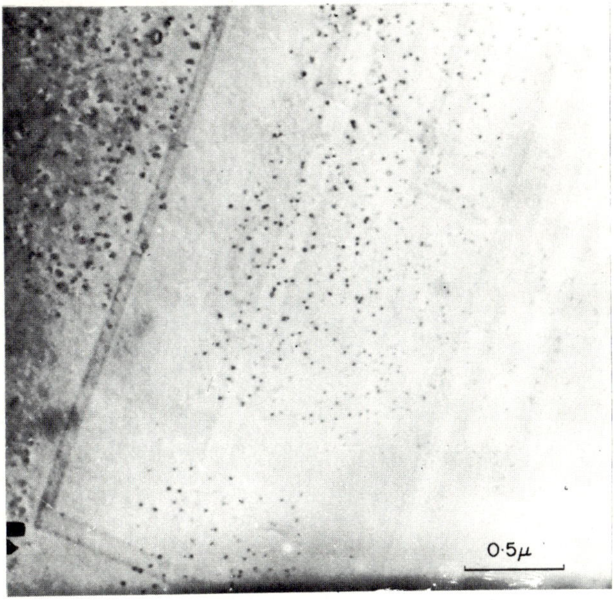

(a)

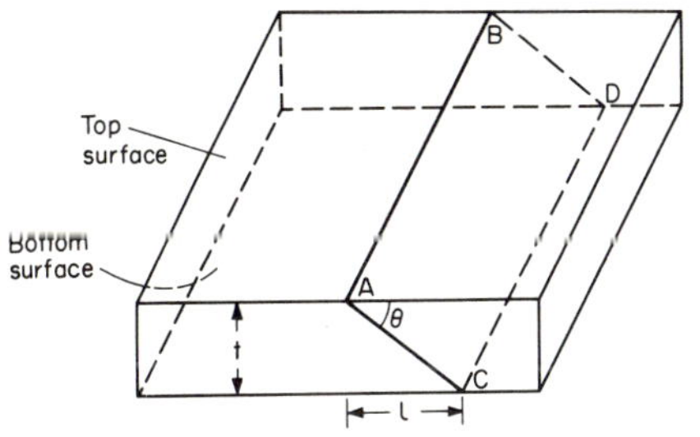

(b)

FIG. 87. (a) Electron micrograph of quenched aluminium showing slip traces, cross-slip traces and small dislocation loops. (b) Diagram showing the relation between the projected width of the slip trace and the foil thickness.

referring to Fig. 88, suppose we have a 211 reciprocal lattice section* and that a slip trace is present on the micrograph parallel to, say, the $[0\bar{1}1]$ direction. Transferring this intersection to the stereogram, *X* in Fig. 88, all the possible slip

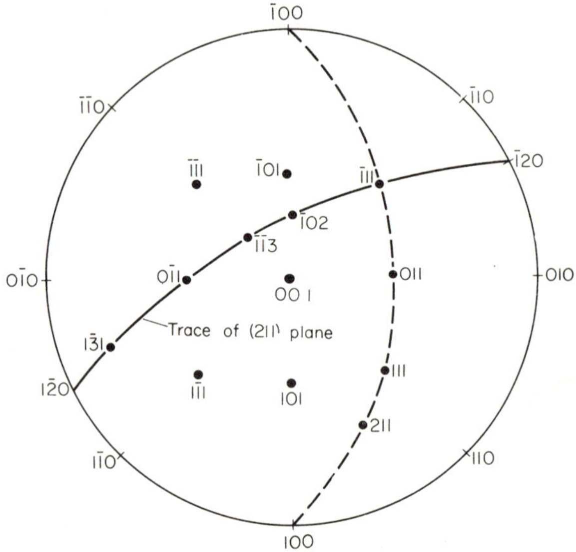

FIG. 88. Standard stereographic projection for a cubic crystal showing how the indices of a 211 reciprocal lattice section are checked. Reflections which can operate will have indices of the poles lying on the trace of the (211) plane.

planes have poles lying on the great circle drawn 90° from *X*. This great circle is indicated by a dotted line in Fig. 88 from which it can be seen that possible slip planes are (100), (111),

*The line being indexed is really a projection in the reciprocal lattice plane of the intersection of the slip plane with the foil plane. The fact that it is not the intersection itself leads to a significant error when the foil is steeply inclined to the electron beam. In such a case, the true foil plane must be determined (by S.A.D. when the foil is normal to the beam) and the projected intersection rotated on the stereogram back into this plane.

(011) and $(\bar{1}11)$. Of the possibilities, one can usually eliminate all but one by using a knowledge of the foil thickness and measuring the projected width of the slip plane, l, as shown in Fig. 87(b). The slip plane *ABCD* is inclined to the foil by an angle $\theta = \tan^{-1}t/l$ which can be measured on the stereogram, from the reciprocal lattice plane normal, along the dotted great circle.

For an extended dislocation the slip plane is delineated by a stacking fault in which case the fringes (see later) run parallel to the line of intersection of the foil surface and the stacking fault plane. For thick foils, the thickness is simply the product of the extinction distance and the number of fringes.

The crystallography of defects which do not intersect the foil surface, such as precipitates bearing an orientation relationship to the matrix and stacking fault tetrahedra, can also be studied as above. In such a case, however, it must be remembered that an edge of the defect lies parallel to an unknown direction in the plane normal to the micrograph, whose trace on the micrograph is delineated by the projected edge.

CONTRAST FROM DISLOCATIONS AND THE g.b ANALYSIS

Image contrast from crystal imperfections arises from the additional phase factor $\alpha = 2\pi\mathbf{g}.\mathbf{R}$ in the equation for the diffraction of electrons by crystals. In the case of dislocations, the displacement **R** giving rise to the phase factor, and hence to the contrast, is simply the Burgers vector; atoms near the core of a dislocation are displaced parallel to **b**. To examine this more closely, consider the lattice surrounding an edge dislocation, Fig. 89(a). Planes near the core are tilted towards the Bragg reflecting condition on one side of the core and away from it on the other side (which side reflects will depend on the orientation of the dislocation with respect to the incident beam). The important consequence of this feature is that the image of the dislocation lies to one side of its true

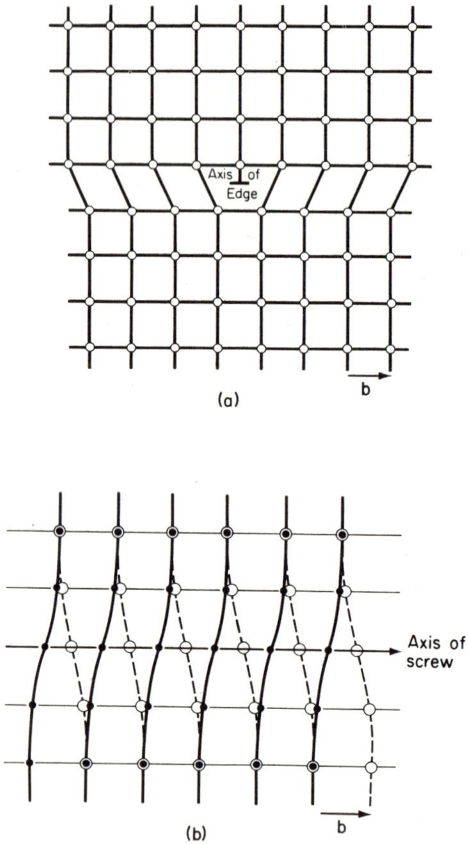

FIG. 89. Diagram showing the tilting of lattice planes around (a) an edge and (b) a screw dislocation. The tilting is in the direction of the Burgers vector **b**.

position, the side tilted towards Bragg reflection. The image will, of course, change from one side of the dislocation to the other if the local orientation of the foil changes as, for example, when a bend contour passes through it (see Fig. 90).

The phase angle introduced by a dislocation is zero when $\mathbf{g} \cdot \mathbf{b} = 0$, and hence there is no contrast, i.e. the dislocation is invisible. Since the scalar product $\mathbf{g} \cdot \mathbf{R}$ is equal to $gR \cos \theta$,

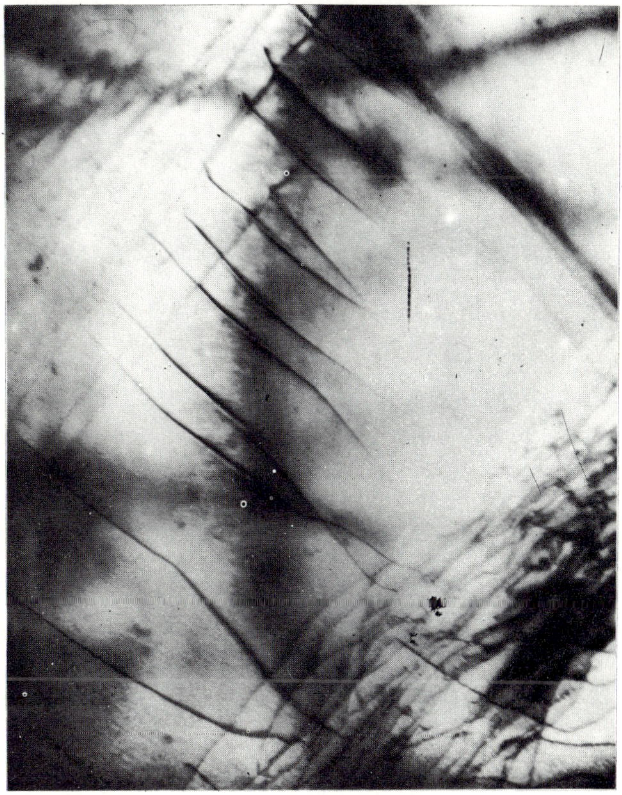

FIG. 90. Electron micrograph showing how the dislocation image lies to one side of the core of the dislocation. The image lies on opposite sides of the core on crossing a bend contour.

where θ is the angle between g and R, then $\mathbf{g}.\mathbf{R}=0$ when the displacement vector **R** is normal to **g**, i.e. parallel to the reflecting plane producing the image.

That this situation should occur when the Burgers vector is parallel to the reflecting planes can be seen from Fig. 89 if the diagrams are turned through 90°. If we think of the lattice planes which reflect the electrons as mirrors, it is easy to understand that no contrast results when $\mathbf{g}.\mathbf{b}=0$, because the dislocation displacement vector **b** merely moves the reflecting planes parallel to themselves, as shown in Fig. 91(a),

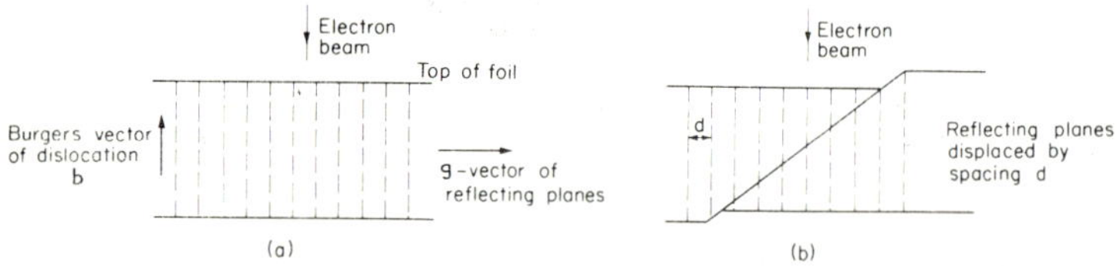

FIG. 91. Diagram to show there is no contrast from (a) a dislocation when $\mathbf{g}.\mathbf{b}=0$, and (b) a stacking fault when $\mathbf{g}.\mathbf{R}=n$.

without altering the intensity scattered from them. Clearly, only displacements which have a component perpendicular to the reflecting plane will produce contrast. A screw dislocation only produces atomic displacements in the direction of its Burgers vector and hence will be completely "invisible" when b lies in the reflecting plane producing the image. For a pure edge dislocation, however, there are some minor atomic displacements perpendicular to **b**. An edge dislocation is, therefore, not completely invisible when **b** lies in the reflecting planes. To illustrate this lost contrast from edge dislocations when $\mathbf{g}.\mathbf{b}=0$, let us examine the contrast from concentric dislocation loops found in quenched Al – Mg alloys, shown in Fig. 92. The inner and outer set of loops have orthogonal Burgers vectors (see Fig. 93) lying in orthogonal $\{100\}$ planes. When the Burgers vector corresponding to one of these sets of concentric loops is made to lie in the

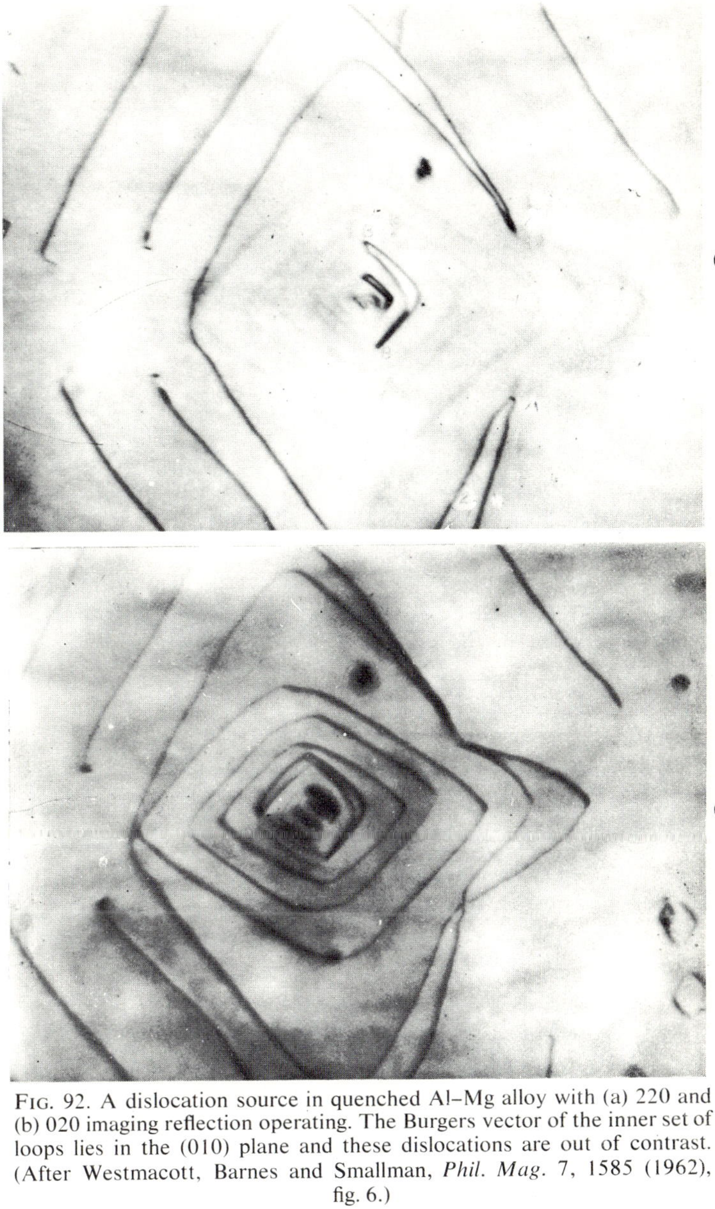

(a)

(b)

FIG. 92. A dislocation source in quenched Al–Mg alloy with (a) 220 and (b) 020 imaging reflection operating. The Burgers vector of the inner set of loops lies in the (010) plane and these dislocations are out of contrast. (After Westmacott, Barnes and Smallman, *Phil. Mag.* 7, 1585 (1962), fig. 6.)

reflecting planes producing the image contrast (e.g., a 020 reflection) the dislocation contrast will virtually vanish since $\mathbf{g.b} = 0$. However, because the loops are predominantly edge in character, the small displacements normal to **b** will not satisfy the $\mathbf{g.R} = 0$ criterion and hence will give rise to

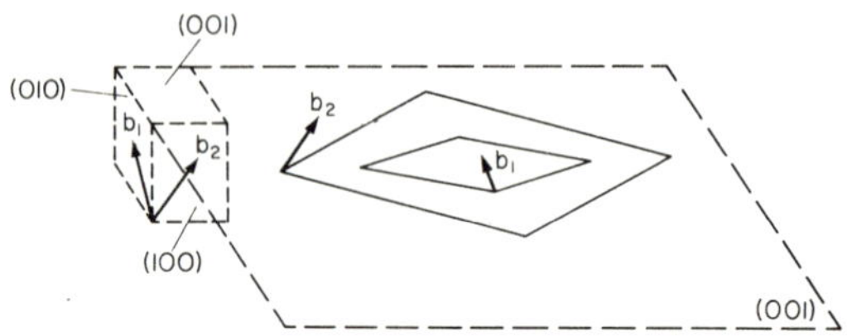

FIG. 93. Schematic diagram showing the orthogonal relationship between the two Burgers vectors $\mathbf{b}_1$ and $\mathbf{b}_2$.

some weak residual contrast. In general, however, a dislocation goes out of contrast when the reflecting plane operating contains its Burgers vector, and this fact is commonly used to determine the Burgers vectors. To establish **b** uniquely, it is necessary to tilt the foil so that the dislocation disappears on at least two different reflections. The Burgers vector must then be parallel to the direction which is common to these two reflecting planes. The magnitude of **b** is usually the repeat distance in this direction.

The use of the $\mathbf{g.b} = 0$ criterion is illustrated in Fig. 94. The helices shown in this micrograph have formed by the condensation of vacancies onto screw dislocations having their Burgers vector **b** parallel to the axis of the helix. Comparison of the two pictures in (a) and (b) shows that the effect of tilting the specimen, and hence changing the reflecting plane, is to make the long helix *B* in (a) disappear in (b). From the diffraction pattern shown in the insert it can be seen that the dislocation *B* is invisible when it has its Burgers vector **b** normal to the **g**-vector of the strong reflecting plane producing the image, i.e., $\mathbf{g.b} = 0$ for this helix.

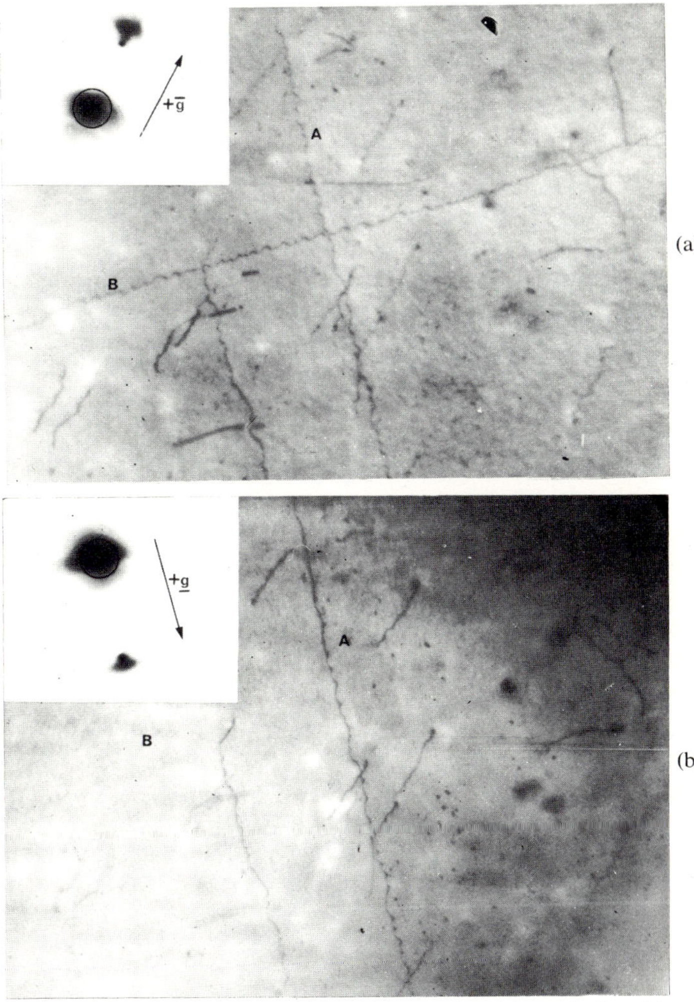

FIG. 94. Electron micrographs to illustrate the **g.b** analysis with helical dislocations for which the Burgers vector lies along the axis of the helix. In (a) the reflecting plane, which is normal to both **g** and the plane of the micrograph (see diffraction pattern inset), does not contain either of the Burgers vectors, but in (b) the Burgers vector of one of the helical dislocations now lies in the reflecting plane (see diffraction pattern inset) and hence this helix is "invisible". (After Hirsch, Howie and Whelan, *Phil. Trans. Roy. Soc.* A, **252**, 499, 1960.)

At this point it should be stressed that the invisibility of a dislocation when the operating reflecting plane contains its Burgers vector does not apply to partial dislocations. The reason for this is that, unlike a unit dislocation, a partial dislocation does not have perfect crystal on both sides of it; one side is joined to a stacking fault.

STACKING FAULT CONTRAST AND THE DISPLACEMENT VECTOR

Contrast at a stacking fault arises because such a defect displaces the reflecting planes relative to each other, above and below the fault plane, as illustrated in Fig. 95. In general,

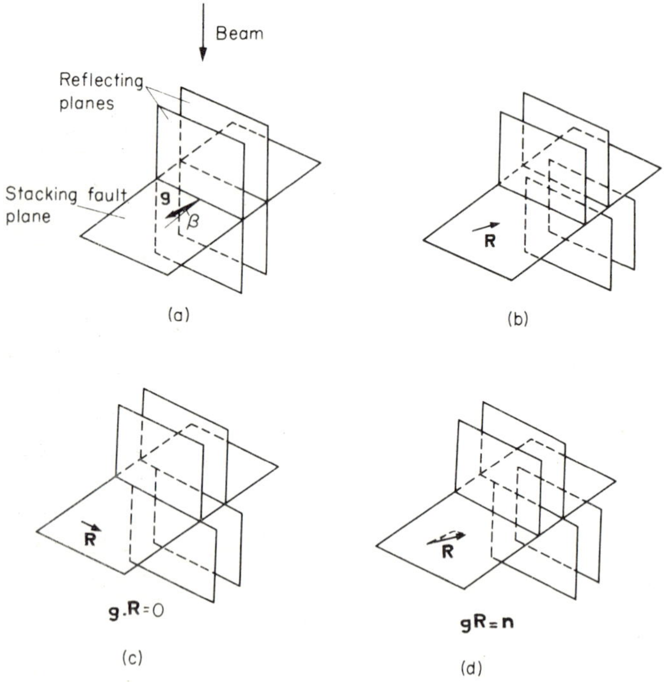

FIG. 95. Diagram showing the displacement of reflecting planes by a stacking fault to give contrast in (b) but not in (c) or (d).

the contrast from a stacking fault will not be uniformly bright or dark as would be the case if it were parallel to the foil surface, but in the form of interference fringes, since the fault usually lies on an inclined plane. The fringes, running parallel to the intersection of the foil surface with the plane containing the fault, appear because the diffracted intensity oscillates with depth in the crystal (see p. 142). Stacking fault fringes can be seen in Figs. 96 and 97. Similar fringe contrast would arise from a wedge-shaped crystal or from a grain boundary inclined to the electron beam.

A useful consequence of the appearance of fringe contrast is that the number of fringes n may be used to estimate the foil thickness and thus aid the determination of the stacking fault plane. Clearly, if the extinction distance t_0 is known, the foil thickness is just nt_0. A source of error in this estimate arises, however, from the fact that the extinction distance varies with the deviation from the exact Bragg reflecting position.

In the formal dynamical theory of electron diffraction at stacking faults, the stacking fault displacement vector **R** is defined as the shear parallel to the fault of the portion of crystal below the fault relative to that above the fault which is regarded as fixed. This vector **R** gives rise to a phase difference $\alpha = 2\pi \mathbf{g} . \mathbf{R}$ in the electron waves diffracted from either side of the fault. The following two important points emerge from the theory. The first point is that stacking fault contrast is absent with reflections for which $\alpha = 2n\pi$, i.e., for which $\mathbf{g} . \mathbf{R} = n$. This is equivalent to the $\mathbf{g} . \mathbf{b} = 0$ criterion for dislocations and can be used to deduce **R**.

The invisibility of stacking fault contrast when $\mathbf{g} . \mathbf{R} = 0$, Fig. 95(c), is exactly analogous to that of a dislocation when $\mathbf{g} . \mathbf{b} = 0$, namely that the displacement vector is parallel to the reflecting planes. The invisibility when $\mathbf{g} . \mathbf{R} = 1, 2, 3 \ldots$ occurs because in these cases the vector **R** moves the imaging reflecting planes normal to themselves by a distance equal to the spacing between the planes. From Fig. 95(d) it can be seen that for this condition the reflecting planes are once again in register on either side of the fault and, as a consequence, there is no interference between waves from the crystal above and below the fault. (It should be noted that, since $\mathbf{R} < \mathbf{b}$, the Burgers vector of the whole dislocation, $\mathbf{g}\,.\,\mathbf{R} = n$ where $n \neq 0$, will occur with high index reflection planes (small d-spacing). Values of $n > 2$ are rarely observed.) The absence of stacking fault contrast when $\mathbf{g}\,.\,\mathbf{R} = n$ is illustrated in Fig. 96(a) and (b) for faults in uranium dioxide.

The second important point that emerges from the contrast theory is that the bright- and dark-field images of a stacking fault are pseudo-complementary at the intersection with the lower surface of the foil and similar at the intersection with the upper surface. Thus, from a comparison of the two images, it is possible to determine which end of a stacking fault intersects the top surface. It also follows from theory that the nature of the first fringe at the top surface of the foil depends only on the sign of α; if the fringe is light on a positive print of the bright field image α is positive; if the fringe is dark, α is negative. Using this information, the sense of $\mathbf{R}$ can be determined. As an example of this, the bright-field and dark-field images of a stacking fault in Cu + 7% Al alloy are reproduced in Fig. 97(a) and (b) respectively. Since the images are pseudo-complementary at Q, this must be the bottom surface of the foil, and since the first fringe at the top on the bright-field micrograph is light, α is positive. Combining this information with a knowledge of $\mathbf{g}$, the sense of $\mathbf{R}$ can be determined ($\alpha = 2\pi\mathbf{g}\,.\,\mathbf{R}$) and, in the case of f.c.c. metals, this indicates the nature of the fault; $\mathbf{R}$ is positive for an intrinsic fault and negative for an extrinsic one.

MEASUREMENT OF THE STACKING FAULT ENERGY

The stacking fault energy, γ, of a metal determines the equilibrium ribbon width of extended dislocations and hence the ease with which such dislocations can climb or cross-slip.

(a) (b)

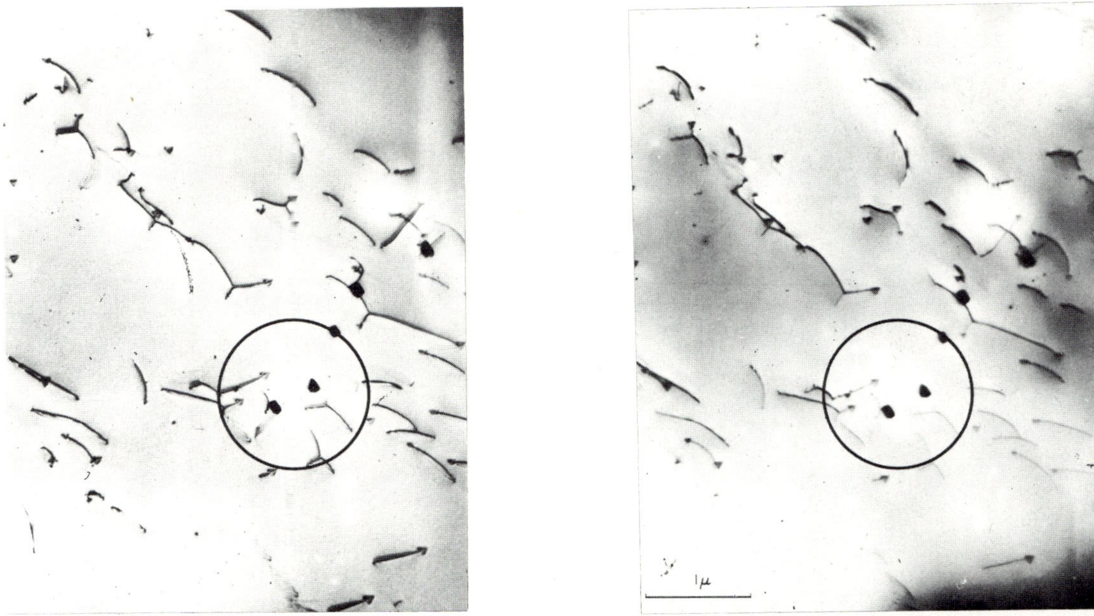

FIG. 96. Stacking faults in UO_2. The faults encircled are (a) in contrast with $\mathbf{g} = 0\bar{2}2$ and (b) out of contrast with $\mathbf{g} = \bar{1}13$. (From K. H G. Ashbee, Stacking faults in uranium dioxide, *Proc. Roy. Soc.* A, **280,** 37 (1964), figs. 2a and 2d.)

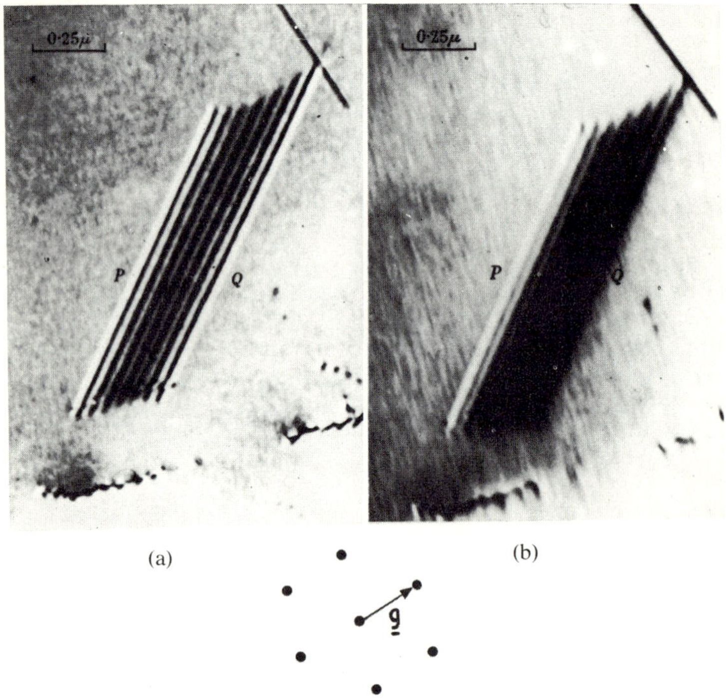

FIG. 97. Contrast from Stacking faults in copper-aluminium alloy, (a) bright-field and (b) dark-field micrograph. (After Hashimoto, Howie and Whelan, *Proc. Roy. Soc.* A, **269**, 1962. fig. 14a and b.)

Since the cross-slip and climb processes are rate-controlling in several important deformation phenomena (e.g., work-hardening, creep, fatigue, etc.), direct measurements of γ are of considerable importance.

In principle, the stacking fault energy can be deduced from the partial separation, *d.* Referring to Fig. 98 which illustrates the case of a mixed extended dislocation, i.e., not pure edge or pure screw, it can be shown that

$$d = \frac{\mu b_p^2}{8\gamma}\frac{(2-\nu)}{(1-\nu)}\left\{1 - \frac{2\nu}{(2-\nu)}\cos 2\alpha\right\} \qquad (9.3)$$

where α is the angle between the total Burgers vector and the ribbon of stacking fault, and ν is Poisson's ratio. Thus, measurement of d on an electron micrograph and substitution in equation (9.1) yields a value of γ. Equation (9.3) shows that the ratio of the width of extended edge and screw dislocations expected is

$$\frac{d_{\text{edge}}}{d_{\text{screw}}} \simeq \frac{7}{3}$$

since α is 90° for an edge and zero for a screw.

In practice, however, the stacking fault energy cannot be deduced from the width of a dislocation observed to be ex-

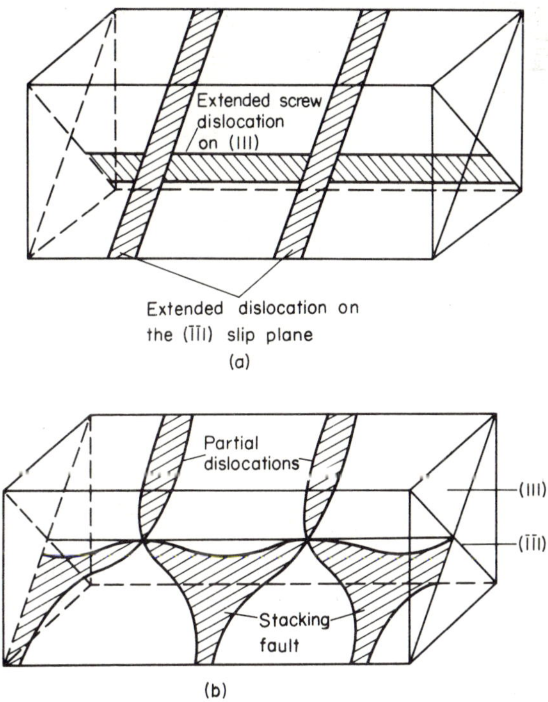

FIG. 98. Diagram showing the formation of extended and contracted nodes by the interaction of extended dislocations on intersecting slip planes.

tended in the microscope, since this separation is not normally the equilibrium separation because of the effect of local stresses. However, γ may be determined by examining stacking faults under near equilibrium conditions. Such conditions are usually found in a hexagonal dislocation network lying in a slip plane formed, as shown in Fig. 98, as a result of the interaction of dislocations with Burgers vectors in this plane. In a metal of low stacking fault energy the dislocations are visibly extended, and the network contains alternate extended and contracted nodes, as seen in Fig. 99. Since all the

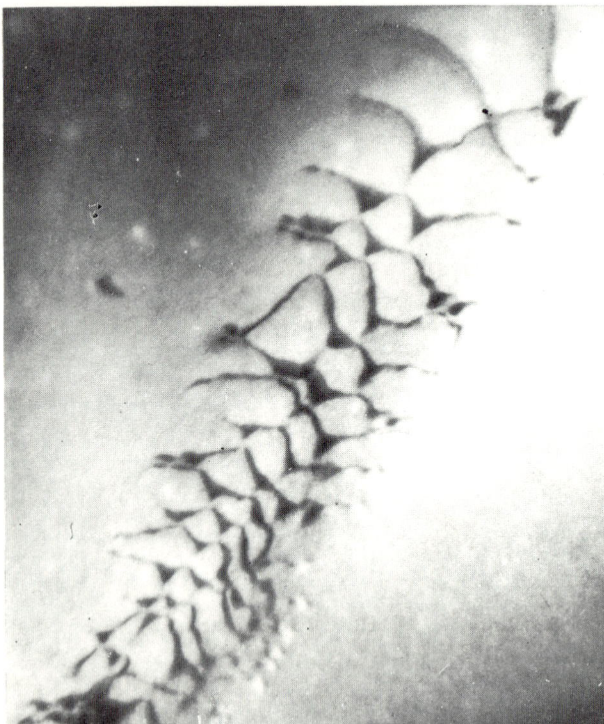

FIG. 99. Micrograph showing extended and contracted nodes in a thin-foil of stainless steel. (After M. J. Whelan, *Proc. Roy. Soc.* A, **249**, 114, fig. 19.)

partial dislocations with vector $\mathbf{b}_p$ are glissile in the plane of the network, the radius of curvature R under the action of the line tension T and the attractive force γ per unit length due to the stacking fault (of energy γ per unit area) can be described by the relation $\gamma = T/R$. The line tension T is given by the relation

$$T = \left[\frac{\mu \mathbf{b}_p{}^2}{4\pi K}\right] \ln (R/r_0)$$

where K is a constant equal to unity for a screw and $(1-\nu)$ for an edge dislocation respectively, μ is the shear modulus, ν is Poisson's ratio and $r_0(\sim \mathbf{b})$ is the dislocation core radius. However, values of γ derived using the measured values of R and the computed values of T are in error because no account is taken of the interaction energy of the nearby partial dislocations and of the variation of line tension along the curved dislocations. These corrections have been estimated and show that γ-values calculated directly from the node radii are too low by as much as 50 per cent. The most reliable values obtained so far are given in Fig. 100 for several copper-based alloys.

The method is limited to materials with $\gamma/\mu\mathbf{b} \lesssim 2 \times 10^{-3}$, and hence excludes most of nominally pure f.c.c. metals. It is possible, however, to extrapolate the curves shown in Fig. 100 to zero solute content to determine a value of γ for the pure metal. The most reliable γ-values obtained in this way are 25, 70 and 225 ergs/cm^2 for silver, copper and nickel respectively.

Other indirect methods are available for determining γ, which depend on the onset of the third stage of work-hardening in single crystals of f.c.c. metals. This stage is associated with the occurrence of thermally activated cross-slip and hence with the stacking fault energy of the metal. Values of γ for several f.c.c. pure metals have also been determined from the texture exhibited after heavy rolling, since this also depends on cross-slip, as discussed in Chapter VI.

NATURE OF PRISMATIC DISLOCATION LOOPS

It is observed using transmission electron microscopy that prismatic dislocation loops are formed as a result of deformation, quenching or irradiation. Metals quenched from a temperature close to the melting point, T_M, contain dislocation loops because the equilibrium vacancy concentration in a metal decreases exponentially with decreasing temperature (from about 10^{-4} at T_M to 10^{-14} at room temperature) and hence the vast majority of the vacancies which exist at high temperatures can be frozen into the lattice by quenching. On

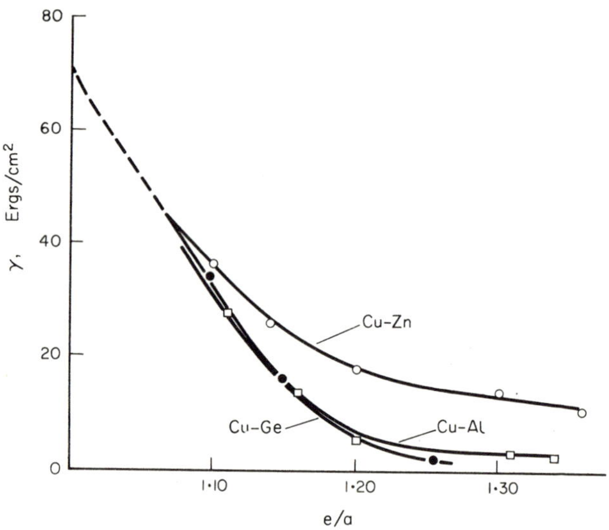

FIG. 100. Variation of stacking fault energy (γ ergs/cm²) with composition (e/a ratio) for several copper-based alloys. (After Thornton, Hirsch and Mitchell, *Phil. Mag.* **7**, 1349, 1962.)

ageing, the excess vacancies migrate and cluster together to form monolayer cavities which subsequently collapse to form dislocation loops. Such loops were first observed in quenched aluminium, but they have since been observed in other metals of different crystal structure as well as in non-metals. Figure 101 shows dislocation loops formed from clusters of vacancies in quenched aluminium. The total vacancy concentration retained by the quenching treatment may be deduced from this micrograph by estimating the size and number of the loops in a given volume (i.e., in a given area of a foil of known thickness) by using the formula $c = \pi r^2 \mathbf{b} n/t$ where n is the number of loops per cm^2, r the loop radius, **b** the Burgers vector and t the foil thickness. The activation energy for the formation of vacancies E_f can therefore be calculated, from the relation $c = \exp[-E_f/KT]$.

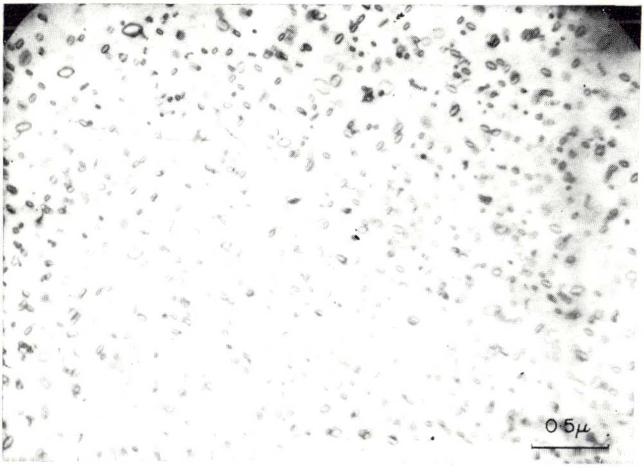

FIG. 101. Electron micrograph showing small dislocation loops in quenched aluminium-alloy. Magnification × 20000.

Dislocation loops formed from clusters of vacancies and interstitials have also been observed in irradiated metals (see Fig. 103) and the concentration of point defects estimated as a function of irradiation dose.

To understand the phenomena of quenching, irradiation and cold work, it is necessary to establish experimentally whether the loops are the result of the aggregation of vacancies or of interstitial atoms. After irradiation or quenching, for example, a sessile dislocation loop in a f.c.c. lattice formed by the condensation of a monolayer of vacancies on a (111) plane is transformed into a prismatic dislocation according to the reaction

$$\frac{a}{3}\left[111\right]+\frac{a}{6}\left[\bar{2}11\right]\rightarrow\frac{a}{2}\left[011\right]$$

The sessile loop formed by the condensation of an interstitial atomic layer on a (111) plane after irradiation is transformed by the reaction

$$\frac{a}{3}\left[\bar{1}\bar{1}\bar{1}\right]+\frac{a}{6}\left[11\bar{2}\right]+\frac{a}{6}\left[1\bar{2}1\right]\rightarrow\frac{a}{6}\left[0\bar{1}\bar{1}\right]$$

the Burgers vector being opposite in sense to that formed from vacancies. Similarly, on cold-working a metal, dislocation loops corresponding to a sheet of vacancies or an extra sheet of ions are formed dependent on the direction of motion of a jog in a screw dislocation.

When the loops can be resolved, the sense of the Burgers vector, and hence the nature of the loop, can be determined by finding out which side of the dislocation its image lies under known diffracting conditions. This follows because the rotations of the lattice planes in the region of the dislocation are in the opposite sense for similarly oriented vacancy and interstitial loops, and therefore the position of the image (outside or inside the true position of the loop) will be different for the two cases.

The position of the image of a dislocation depends on the sign of $(\mathbf{g}\,.\,\mathbf{b})\,\mathbf{s}$, where $\mathbf{s}$ is the vector in reciprocal space defining a departure from exact diffraction conditions. A change in sign of any one of these vectors results in the contrast moving to the other side of the dislocation line (see Fig. 90). The observed consequence of the image of a dislocation loop moving from one side of the dislocation to the other is that the apparent size of the loop changes.

In some practical cases this apparent change in size has been noted on either side of an extinction contour, i.e., changing the sign of **s** by tilting the specimen but keeping the sign of **g** constant. The sign of **s** is determined from the positions of Kikuchi lines (see a previous section, p. 159). However in metal foils, dislocations are usually studied in regions of good transmission, which are found on either side of dark bands corresponding to high absorption and associated with a pair of low order Bragg reflections +**g** and −**g** (see Fig. 81). The sign of **s** is positive on either side of the dark band, but the sign of the strong reflection **g** changes, so that if the same loop is observed on either side of the band in turn (by tilting the specimen), there is a change in the sign of (**g** . **b**)**s**.

By noting the change in size of a particular loop when the sign of (**g** . **b**)**s** is changed, one can decide between a similarly oriented vacancy or interstitial loop, for example (a) or (b) in Fig. 102. It is evident that the lattice rotations, and hence the Burgers vector, are in the opposite sense and (a) would appear to decrease in size and (b) would increase in size as (**g** . **b**)**s** becomes positive. However, the lattice rotations (or sense of Burgers vector) of a vacancy loop on one plane are identical

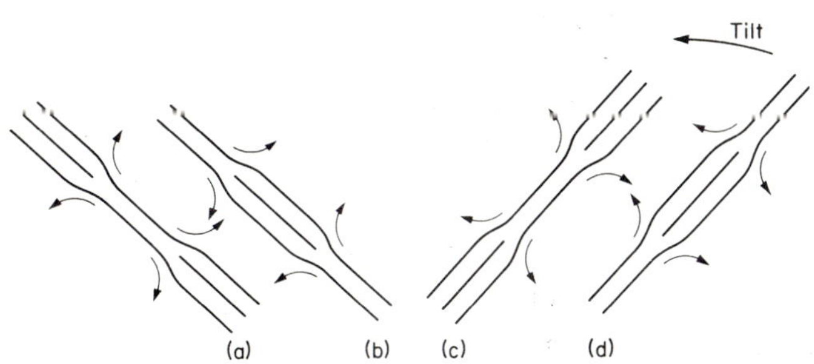

FIG. 102. Diagram showing the tilting of the lattice planes for vacancy- and interstitial-type loops.

to those associated with an interstitial loop on an orthogonally inclined plane (cf. loops (a) and (d) in Fig. 102) and vice versa (cf. loops (b) and (c) in Fig. 102).

After studying the apparent size of the loop under positive and negative reflecting conditions, one is then faced with two alternative possibilities, namely a vacancy loop on one plane or an interstitial loop on an orthogonal plane equally inclined to the foil surface. It is possible to decide between these loops, however, on simple geometric grounds by tilting the foil, not by a degree or so, as in the normal stereo stage, but through a large angle ($\sim 20°$). The projected area of the dislocation loop changes in the opposite sense for the vacancy and interstitial loop respectively. Thus, if the tilt is carried out about an axis parallel to the major axis of the loop, i.e., in the direction of the arrow in Fig. 102, an interstitial loop of orientation (b) will increase in size, while a vacancy loop of orientation (c) will decrease in size. To carry out such a large angle tilt about a given tilt axis, it is necessary to use a special goniometer stage of the type shown in Fig. 71 for holding the specimen in the microscope.

It must be remembered that to obtain the correct sense of **g** it is necessary to orient the diffraction pattern correctly with respect to the corresponding micrograph. To do this one must correct not only for the rotation introduced on switching from micrograph to diffraction conditions, i.e., when the intermediate lens current is reduced so as to focus this lens on the back focal plane of the objective lens, but also for the extra inversion which the specimen image suffers in the objective lens. These rotations are discussed in detail in Chapter VII.

Figure 103 shows the results of this type of analysis carried out on aluminium bombarded with 38 MeVα-particles for dislocation loops in the vicinity of the stopped α-particles. The loops generally lie on $\{110\}$ planes with a $[110]$ Burgers vector. A comparison of Fig. 103(a) and (b), or (c) and (d) shows the change in size of the loop image when the sign of **g** is changed, and a comparison of (a) and (c) or (b) and (d) shows the change in loop image on tilting 15°. The analysis shows that these loops are interstitial in nature, as previously inferred from observations of the annealing behaviour of the

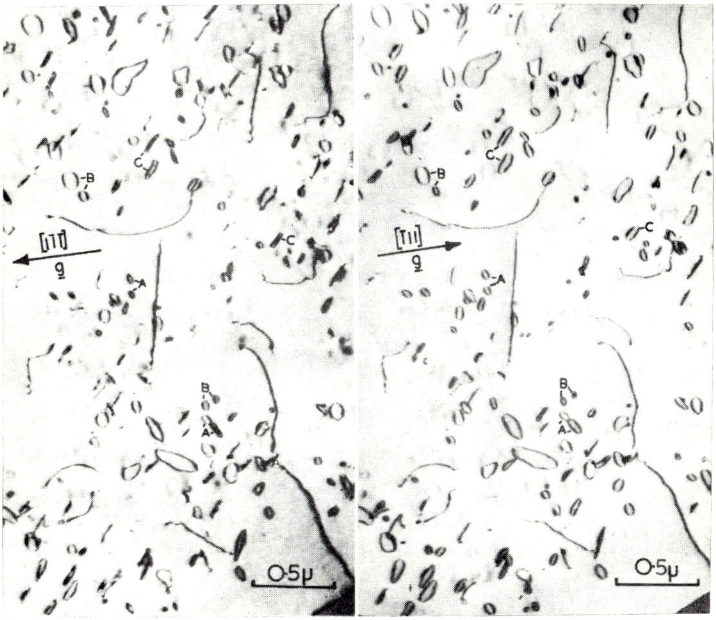

(a) (b)

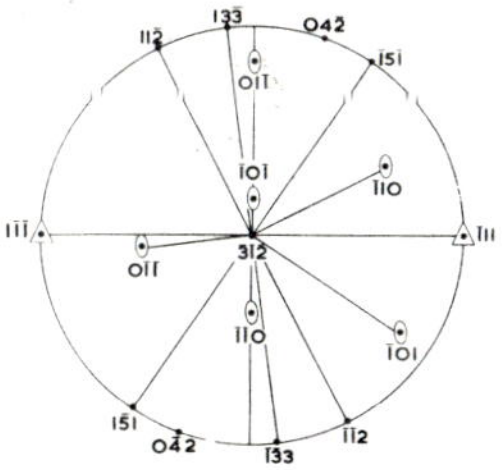

FIG. 103(a, b). Electron micrograph showing the variation in the size of the image with the sign of **g** for dislocation loops in α-irradiated aluminium. (After D. J. Mazey, R. S. Barnes and A. Howie, *Phil. Mag.*, **7**, 1861, 1962, fig. 4.)

loops. Similar studies on quenched metals confirm that the loops are vacancy in nature as expected, since the equilibrium vacancy concentration increases exponentially with increase

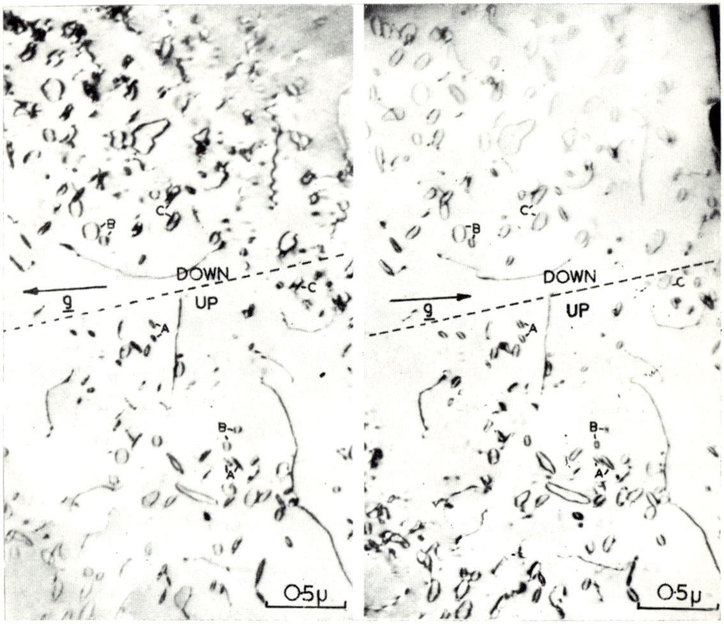

(c) (d)

FIG. 103(c, d). Variation in the size of image with **g** and with large-angle tilt.

in temperature, and hence on quenching an excess vacancy concentration is retained. In cold worked materials both types of loops are observed.

DISLOCATION DENSITIES

The technique of transmission electron microscopy is particularly useful for observing the density and arrangement of dislocations produced by various deformation processes, e.g., cold work, creep, fatigue, etc. Several methods have been used to determine the dislocation density from standard electron micrographs. For low dislocation densities (i.e., ~ 1% strain) ρ is given by the relation $\rho = 2\,n/A$, where n is the number of intersections that the dislocations make with both foil surfaces in an area A. At higher densities (10^{10} lines/cm^2) ρ may be estimated by two alternative methods depending on the particular distribution observed. When the dislocation arrangement shows a cell structure then $\rho = 2n/Lt$, where n is the number of intersections made by the dislocations with a set of random lines of length L drawn in the micrograph and t is the foil thickness. In the absence of a cell structure the average dislocation density is given by $\rho = \frac{1}{t}\left\{(n_1/L_1) + (n_2/L_2)\right\}$ where n_1 and n_2 are the average numbers of intersections made with the dislocations by two orthogonal sets of parallel grid lines of lengths L_1 and L_2 drawn on the micrograph and t is the foil thickness.

Density measurements in a number of metals of different crystals structures show that there is a relation between the resolved shear stress τ and ρ of the form

$$\tau = \tau_0 + \alpha\mu\mathbf{b}\sqrt{\rho},$$

where μ is the shear modulus, **b** the Burgers vector and α a constant of the order of 0·5 at a given temperature. In metals with f.c.c. structure τ_0 is very small or zero, but τ_0 is appreciable in b.c.c. metals (~ 10^3 psi) and may be identified with the lattice friction parameter σ_i deduced from the dependence of the yield stress σ_y on grain-size $2d$, i.e., the Petch equation $\sigma_y = \sigma_i + k_y d^{-1/2}$

PRECIPITATION PHENOMENA

One of the more direct applications of the thin-film electron transmission microscopy technique is in the observation of precipitates or zones (i.e., clusters of solute atoms) in certain alloys. It is well known that the structural changes which take place in precipitation-hardening alloys (e.g., Al–Cu, Cu–Be,

Ni–Al) during the initial stages of ageing are on too fine a scale to be observed in the light microscope, yet it is in these early stages that the most spectacular changes in mechanical properties occur. However, with the electron microscope, the complete precipitation process may be followed from the initial clustering of solute atoms on certain crystallographic planes in the solvent lattice, through the intermediate, or metastable, precipitate stage to the formation of the new precipitate found labelled in the equilibrium diagram. Figure 104(a) shows the precipitates in nickel–aluminium alloy; this system is the basis of the successful creep resistant alloys of the Nimonic series. In this and many other alloy systems the

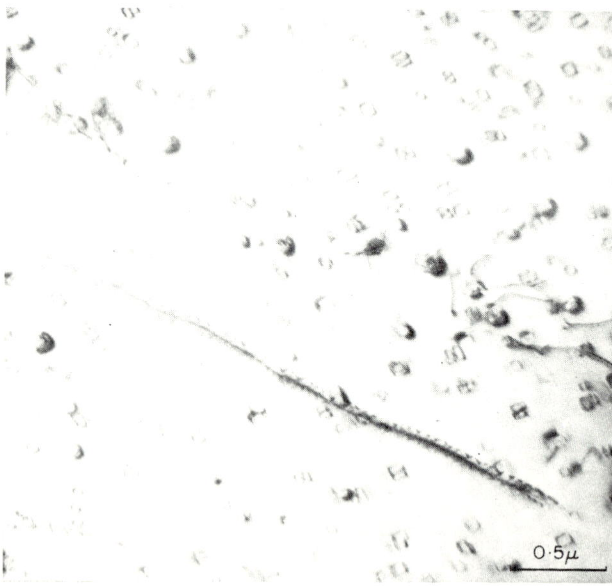

FIG. 104(a). Electron micrograph of γ-precipitate in Ni/5 Wt% Al alloy oil quenched from 1300°C and aged at 645°C for156 hours (unpublished work of Beeston and Smallman).

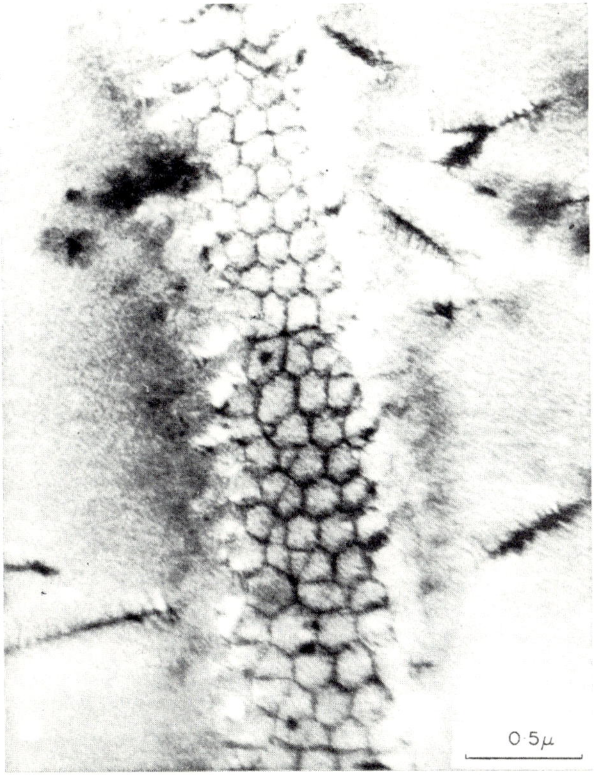

FIG. 104(b). Precipitation on isolated dislocations and finer precipitation on networks in iron (< 0.05%C) quenched from 690°C and aged 3 months at room temperature.

precipitation sequence characteristic of any given ageing temperature can be followed in the microscope and correlated with the accompanying changes in physical and mechanical properties.

In such studies it is observed that precipitation often occurs preferentially at dislocations. It is probable that such precipitates start as an "atmosphere" at the core of the dislocation which begins to grow along certain crystallographic directions after saturating the dislocation. An example of precipitation on dislocations is shown in Fig. 104(b); clearly such observations are of direct importance to the development of the theory of strain-ageing and quench-ageing.

The elastic strains which result from the precipitation of a fully coherent particle in a matrix of different lattice parameter are known to be one source of precipitation-hardening. These strains take the form of atomic displacements **R** around the particle which give rise to diffraction contrast because they introduce a phase angle $\alpha = 2\pi\mathbf{g}\,.\,\mathbf{R}$ (see Chapter VIII). Dark-field microscopy may be used to determine the magnitude of the coherency strain around such a precipitate particle, and also whether the particle is in tension or compression. In the simplest case of a spherical particle of radius r_0, the atomic displacements are radial and given by $R = \epsilon r_0^3/r^2$ in the matrix, i.e., $r > r_0$, and $R = \epsilon r$ within the particle. Because the displacements are radial, those atoms in the reflecting planes producing the electron-optical image are displaced parallel to the reflecting planes and therefore the diffraction conditions in the planes passing through the centre of the particle will be unaltered. As a consequence the image of the particle will be characterized by a "line of no-contrast" through the centre of the image and perpendicular to the reciprocal lattice vector g. This is shown schematically in Fig. 105 and can be seen in

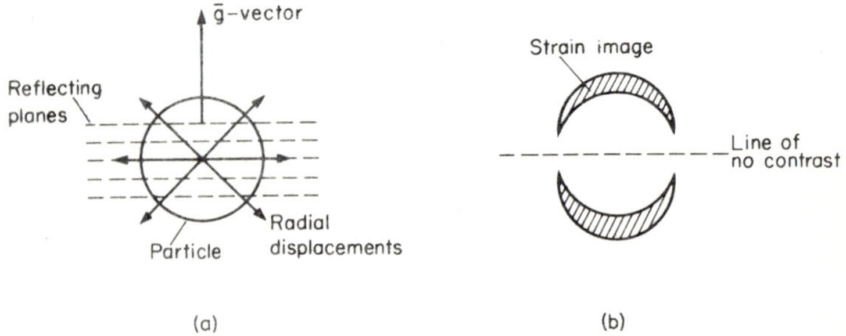

FIG. 105. Diagram to illustrate the strain contrast from a spherical precipitate. When the radial displacements lie in the operating reflecting plane (a) the strain-contrast vanishes (b).

an actual micrograph from a copper-2% cobalt alloy in Fig. 106.

For a particle in the centre of the foil, the bright field image is symmetrical about the line of no contrast. Near the surface of the foil, however, the particles give rise to anomalously wide, asymmetrical images. By measuring the width of a symmetrical image, i.e., in the centre of the foil, and the particle

FIG. 106(a). Bright-field electron micrograph with g = 311, showing precipitates in copper-cobalt alloy.

radius on a given micrograph, the magnitude of the misfit ϵ may be estimated.

The sign of ϵ can be determined from a comparison of the

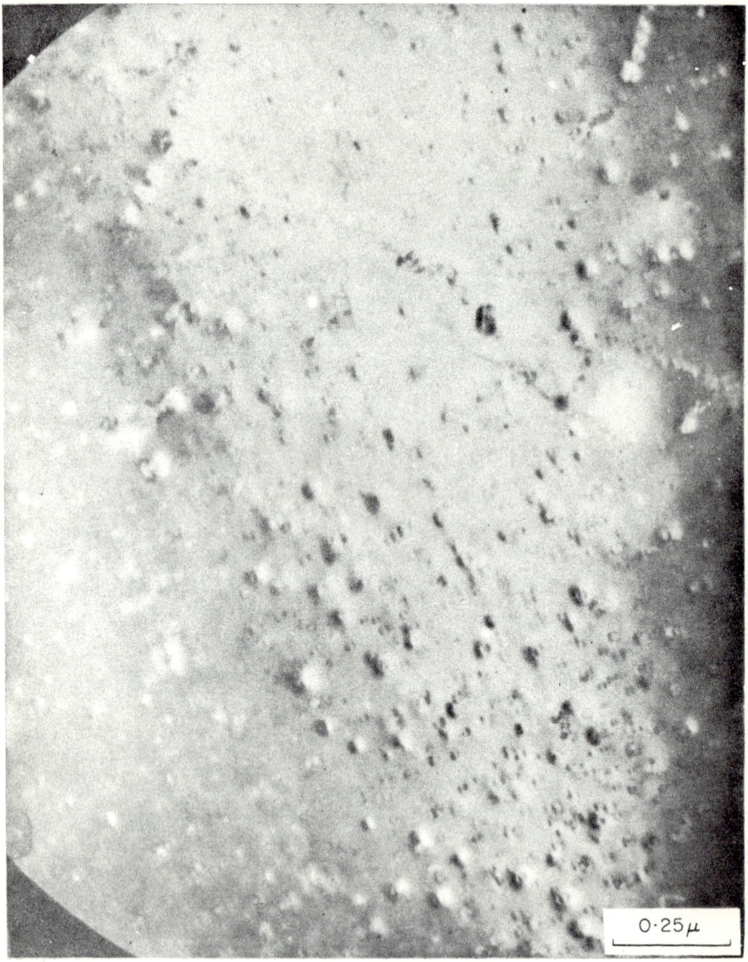

FIG. 106(b). Dark-field micrograph with $\mathbf{g} = 200$, showing precipitates in copper-cobalt alloy. Extensive (anomalous) strain-contrast is observed for particles near the surfaces of the foil. (After Ashby and Brown, *Phil. Mag.*, **8**, 1083, 1963, fig. 13).

bright- and dark-field pictures. Thus, a bright-field micrograph will show opposite symmetry for particles near the top and bottom of the foil, but the effect of anomalous absorption is to give the dark-field image consistent symmetry. In other words, the bright- and dark-field asymmetrical images are complementary at the bottom of the foil but similar at the top of the foil, as shown schematically in Fig. 107; this situation is similar

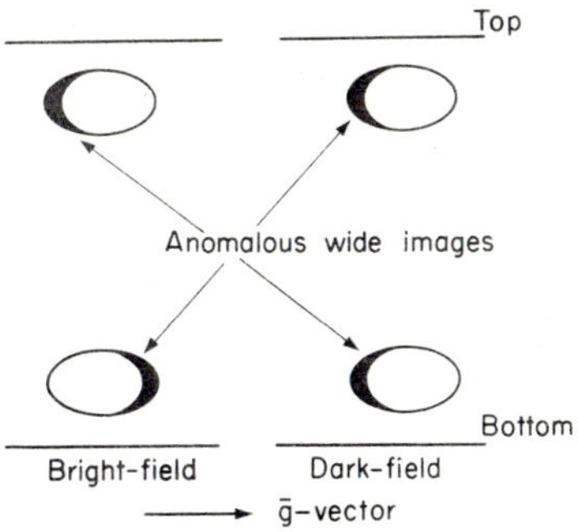

FIG. 107. Schematic diagram showing the relation between the anomalous wide images and the surfaces of the foil in bright and dark-field micrographs.

to that described previously for stacking fault contrast. Figure 106(b) shows that about one image in ten is anomalously wide (i.e., lies near the surface of the foil) and white in the direction of positive **g**. The theory shows that this is consistent with ϵ being negative (i.e., vacancy-type in character) which is consistent with the lattice parameters of cobalt and copper which are 3·54 Å and 3·61 Å respectively.

The technique can, of course, be applied to non-spherical particles and in the case of plate-shaped particles of volume V, r_0^3 would be replaced by $3V/4\pi$. It follows that this technique of studying the asymmetry of the image may also be used to determine the nature of small unresolvable dislocation loops commonly found in irradiated materials, when the more conventional contrast analysis described previously cannot be applied. In this case prismatic loops close to the surface of the foil will show anomalously wide asymmetrical images, and the sense of the asymmetry is determined by the sense of the Burgers vector of the loop.

HEATING-STAGE EXPERIMENTS

Important information of both a qualitative and quantitative nature can be obtained by studying the behaviour of thin metal foils in a heating stage in the microscope. Such structural phenomena as transformations, recrystallization and precipitation are clearly suitable subjects for investigation. Heating-stage observations can also provide quantitative information on dislocation climb and vacancy migration, for example by studying the behaviour of dislocation loops on annealing. The excess resistivity produced by quenching a metal from a high temperature is associated with the vacancies retained in excess of the thermodynamic equilibrium concentration. On heating the sample the resistivity is observed to anneal-out in two stages, the first with an activation energy characteristic of the energy of vacancy migration and the second with that of energy of self-diffusion, i.e., $E_D = E_f + E_m$. Microscope experiments show that the first stage is associated with the migration of individual vacancies to form large clusters, which subsequently collapse to form dislocation loops (i.e., with a process of activation energy E_m) and the second stage is associated with the disappearance of the dislocation loops (i.e., by the continual formation and removal of vacancies from the loop).

The dislocation loops in quenched aluminium, for example, are observed to anneal-out in a few minutes at a temperature of approximately 200°C. The rate of shrinking of a loop depends on the ease with which the dislocation line can create and emit a vacancy (i.e., dislocation climb). To treat this mathematically use is made of the fact that vacancies are also emitted from jogs on dislocations. Thus, the climb process will be controlled by the velocity of a jog. This is given by the expression

$$V = z\nu_a \mathbf{b} \exp(-E_D/KT)[\exp(F_c\mathbf{b}^2/KT) - \exp(F_s\mathbf{b}^2/KT)]$$

where z is the co-ordination number, $\mathbf{b}$ the Burgers vector of the dislocation, ν_a an atomic frequency and E_D the activation energy for self-diffusion. The terms in square brackets represent the opposing force effects on the dislocation. F_c is the force due to applied or local stresses which tend to make the dislocation climb, e.g., by a jog emitting a vacancy, whereas $F_s = KT/\mathbf{b}^2[C/C_0]$ is the "chemical" force due to C, the supersaturation of vacancies relative to the equilibrium concentration C_0, which tends to make the jog accept a vacancy and therefore move in the opposite direction.

In thin foils it is reasonable to neglect F_s, since the vacancy can escape to the surface of the foil, and by substituting for F_c an expression for the force due to the line tension of a circular loop it is possible to show that the variation of loop radius with time t is given by

$$r = r_0(1 - t/\tau)^{1/2} \qquad (9.3)$$

where the lifetime of a loop, initially of radius r_0, is given by

$$\tau \quad \frac{r_0}{\alpha^2 z \nu_a \mathbf{b}} \left[\exp E_D/KT\right]$$

In this expression α is a parameter which depends on the shear modulus, Poisson's ratio and to some degree on temperature, but is independent of the loop radius. Silcox and Whelan have shown that equation (9.3) describes their heating-stage observations on the annealing of loops in aluminium quite well and inserting experimentally determined values in the expression obtain good agreement with the accepted figure for the activation energy of self-diffusion, $E_D = 1{\cdot}3$ eV. Similar experiments have since been carried out on a wide variety of materials, e.g., copper, nickel, zinc, vanadium, niobium, magnesium oxide, and titanium carbide.

Loops containing stacking faults have also been annealed out and a value for the stacking fault energy derived.

COOLING-STAGE EXPERIMENTS

Thin film electron microscopy has proved useful in the study of the products of the shear-type martensitic transformation, such as those found in steels. In this type of study it is possible to follow the progress of the transformation directly in the microscope using a cooling stage, provided a suitable alloy is chosen such that the temperature for the start of the transformation (M_s) is near to room temperature.

Metastable β-brass produced by quenching also transforms martensitically at sub-zero temperatures and has been studied directly in a cooling stage. It is found that for such an alloy the M_s temperature is higher in thin films than for bulk material due to a reduction in the magnitude of surrounding constraints. An alloy containing 61% copper normally transforms to a f.c. orthorhombic structure at −60°C, but transforms spontaneously at 20°C to a f.c.c. structure when the alloy is electrolytically thinned to foils less than about 1500 Å thick. The transformation is similar to normal martensitic transformations but the interface between the martensitic phase and parent phase is largely determined by the thickness of the foils and does not follow any particular habit plane as in the martensitic transformation in bulk materials. Figure 108(a) shows the structure of the martensitic phase; the fine striations are almost certainly produced by deformation twins. Selected area diffraction shows that thin-foils which initially have a cubic orientation, $(001)\beta$, transform in two different ways to produce the f.c.c. structure. In one case the transformation can be accounted for by a pure homogeneous strain, but this only occurs in the very thin regions at the edges of the foil and, in the other, by the combination of a rotation, a homogeneous strain and a half-twinning shear.

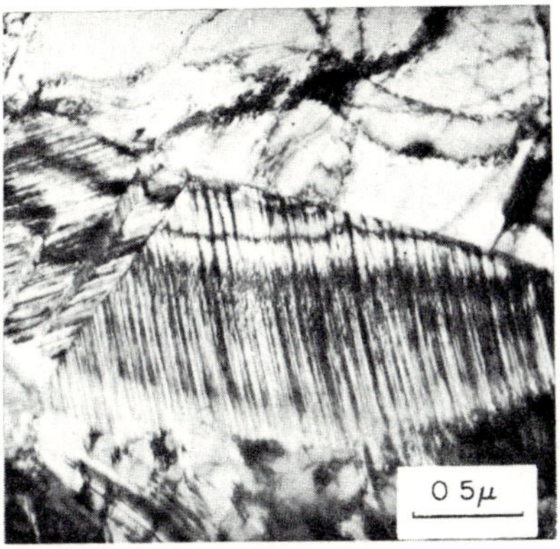

(a)

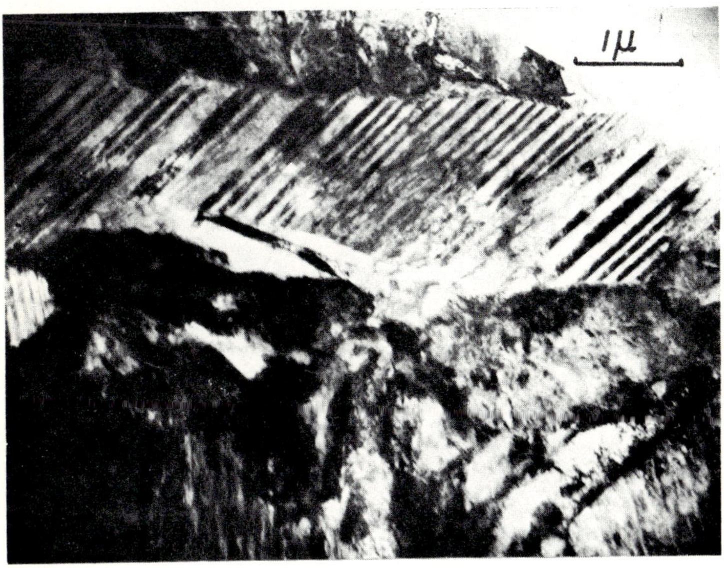

(b)

FIG. 108. (a) Electron micrograph showing martensitic platelets in transformed β-brass. (After D. Hull, *Electron Microscopy and Strength of Crystals*, p. 939, fig. 19, Interscience Publications.) (b) β-NiTi alloy (unpublished work of A. Ball and R. E. Smallman).

Similar work has been carried out on a 30% Ni–Fe alloy which transforms martensitically from f.c.c. to b.c.c. at −30°C. Almost identical effects to those found in β-brass were observed, which indicates that the martensitic transformation in thin foils probably occurs above the M_s of bulk material in all systems. Martensitic transformations occur in a good many β-phase $\frac{3}{2}$ electron compounds and Fig. 108(b) shows an example of martensitic plates in a 50/50 Ni/Ti alloy.

SUGGESTIONS FOR FURTHER READING

G. Thomas and J. Washburn (Editors), *Electron Microscopy and Strength of Crystals*, Interscience, 1963.

H. Hashimoto, A. Howie and M. J. Whelan, *Phil. Mag.*, **5**, 967 (1960).

A. Howie and M. J. Whelan, *Proc. Roy. Soc.* A, **263**, 217 (1961).

M. J. Whelan and P. B. Hirsch, Electron diffraction from crystals containing stacking faults, I, *Phil. Mag.*, **2**, 1121 (1957).

M. J. Whelan and P. B. Hirsch, Electron diffraction from crystals containing stacking faults, II, *Phil. Mag.*, **2**, 1303 (1957).

H. Hashimoto, A. Howie, and M. J. Whelan, Anomalous electron absorption effects in metal foils: theory and comparison with experiment, *Proc. Roy. Soc.*, A, **269**, 80 (1962).

M. F. Ashby and L. M. Brown, On diffraction contrast from inclusions, *Phil. Mag.*, **8**, 1649 (1963).

CHAPTER X

METALLOGRAPHY AT THE ATOMIC LEVEL

FIELD-ION MICROSCOPE

The present-day electron microscopes can easily attain a resolution of 10Å, which is sufficient to resolve the individual molecular structure in non-metallic materials having a lattice spacing of this value. Figure 109 shows the $(20\bar{1})$ planes of spacing 12Å in platinum phthalocyanine; the extra half-plane of platinum ions, characteristic of an edge dislocation, is clearly evident. The lattice spacings of most metals are however about 3Å and, consequently, the individual atoms in the structure cannot be seen. It is possible that the ultimate resolution of microstructure will be achieved with a high-voltage electron microscope, but until this becomes available it is necessary to use the field-ion microscope to resolve the atomic lattice of a metal.

A sectional diagram of a typical field-ion microscope is shown in Fig. 110. The metal specimen usually takes the form of a wire, and the "emission" end is electro-polished to a hemispherical tip with a radius equal to a few hundred Angstroms, while the other end is welded to a filament between tungsten cathodes cooled with liquid and solid nitrogen. The whole assembly is mounted in a glass chamber and, after being evacuated to 10^{-6} mm of Hg, a trace of helium gas is allowed to leak in from another vessel at a pressure of 10^{-3} mm. A fluorescent screen, made up from a copper-activated ZnS.CdS phosphor, is located an inch or so away from the specimen tip and records the atomic image of the specimen as a pattern of

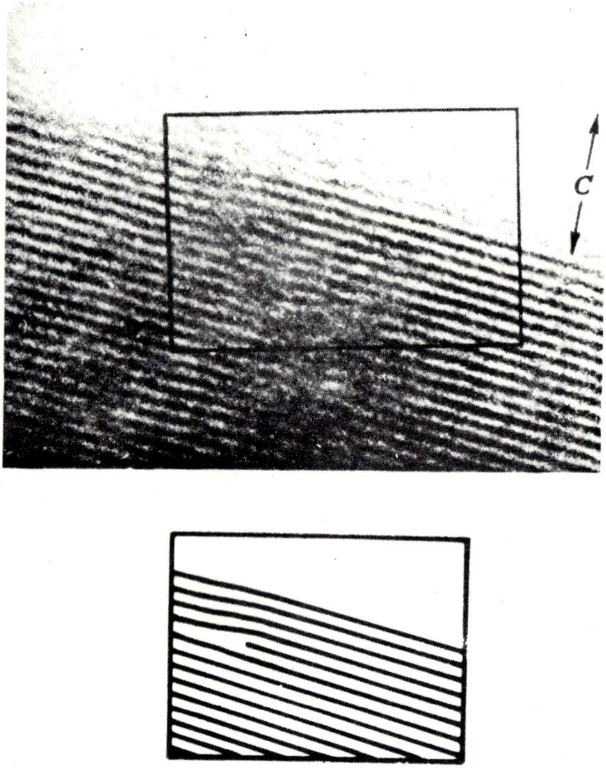

FIG. 109. Electron micrograph showing direct resolution of lattice planes in a thin specimen of platinum thalocyanine. (After Menter, *Proc. Roy. Soc.*, A, **236**, 119, 1956.)

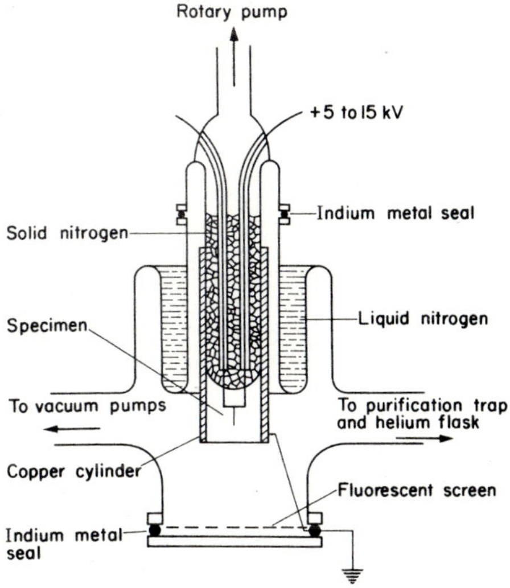

FIG. 110. Schematic diagram of the field-ion microscope. (After A. H. Cottrell, *J. Inst. Metals*, **90**, 1962, p. 450, fig. 1.)

bright spots, each spot being produced by a helium ion, since the specimen is held at a positive potential of about 5 kV relative to the screen.

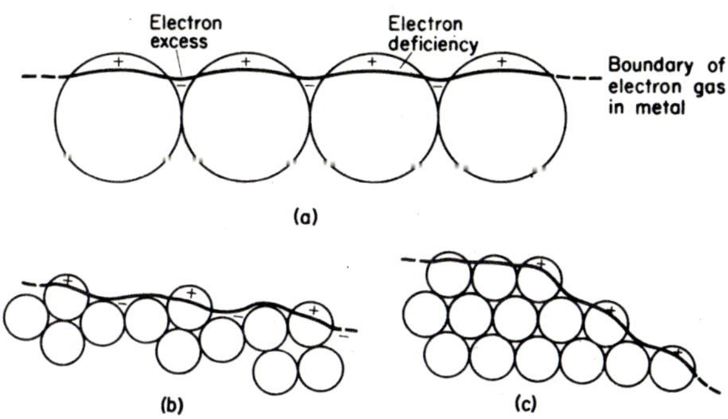

FIG. 111. Schematic diagram of boundary of free-electron gas at the surface of specimen tip, drawn inwards by the applied field; (a) close-packed plane, (b) greater exposure of some positive ions on open-packed planes, (c) edges of crystal planes.

By making the tip positively charged* the free electrons in the metal are pulled inward slightly and, as a result, the positively charged metal ions on the tip surface are partly exposed as shown in Fig. 111(a). The image of these partially exposed ions is then carried to the screen by the gas atoms, because the helium atoms in the vicinity of the surface give up an electron to the metal and hence become positive ions. These gas ions then accelerate down the line of force radiating from the metal ion to the surface and, since these lines are straight and radial, the pattern on the screen faithfully reproduces the pattern of ionization centres on the metal tip. Under the typical conditions mentioned above, a helium ion current of 10^{-9} amps is produced and the total image may be photographed in an exposure time of about a minute.

Although the magnification of the instrument is about 10^6 (ratio of screen distance to tip radius) not all the surface atoms are sufficiently strongly ionized to produce an image. In fact only those atoms exposed above the general surface level become strongly ionized and hence give images on the screen. Such atoms occur where close-packed planes emerge from the surface and examples are shown in Fig. 111(b) and (c).

Clearly, the instrument must be operated at a voltage sufficiently high to produce an image sufficiently intense to be recorded, but not so high that ions of the metal are pulled off the surface. The ability of the metal to resist this "field" evaporation is governed by the cohesive strength of the metal and a serious limitation of the technique at present is that only strongly bound metals such as W, Mo and Pt can be examined. However, it may be possible in future, by using image intensifiers and gases which ionize more easily, to bring the weaker metals within the scope of the instrument.

*In the field emission microscope, the tip is negatively charged and electrons are ejected from the metal with high velocities at right angles to the lines of force, which tends to blur the image and reduce the resolution.

APPLICATIONS

In principle, the instrument is able to show up (a) single parent and impurity atoms, lattice vacancies and interstitial atoms, (b) the atomic structure of grain boundaries and (c) the core structure of a dislocation. Figure 112 shows a typical field-ion image of a tungsten tip. The characteristic dome-like image arises from the fact that the surface of the tip is hemispherical and its intersection with a set of parallel atomic planes gives a stack of approximately concentric discs, one disc for each plane intersected. Intermingled with the dome-like regions are crystalline facets arising from regions where resolvable crystal planes happen to lie parallel to the surface.

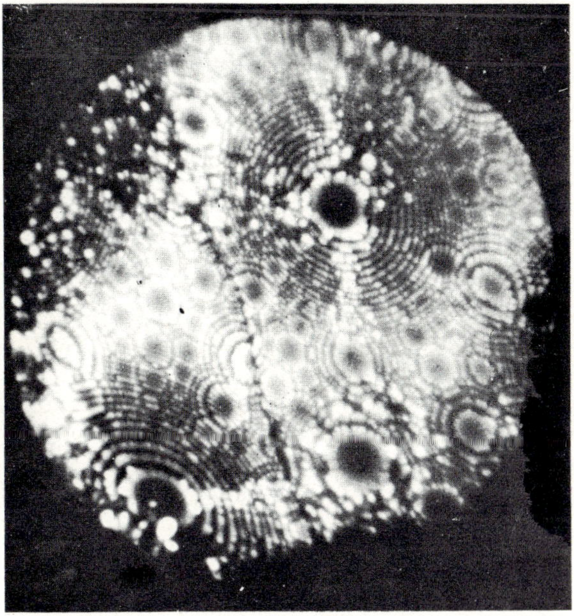

FIG. 112. Field-ion micrograph from pure tungsten, imaged with helium ions at 77°K with a tip radius of 700 Å, showing a grain boundary with 15° misorientation and 122 axis. (After M. S. Wald, *J. Phys. Soc. Japan,* **18**, suppl. II, 324, 1963, fig. 7.)

The figure also shows the radical change of structure across a large-angle boundary but even in the thickest regions some atoms fit both lattices. It is also evident that the thickness of the boundary is not uniform along its length, but varies between one and three atoms in width; this thickness agrees with theoretical predictions about such boundaries.

Single vacancies have been observed by Müller in platinum quenched from just below its melting point as dark holes in lines of brightly visible atoms. Point defects have also been seen in tungsten after bombardment with neutrons. With the field-ion microscope it is possible to bombard a specimen with α-particles, e.g., emitted from a polonium source placed along side it, while it is under observation. It is also possible to strip off successive surface layers by "field" evaporation while the specimen is still under observation and see the distribution of atoms, impurities and lattice defects in the metal. Müller was able to show that in quenched platinum about one lattice site in two thousand was empty, which is consistent with the value of the activation energy for the formation of vacancies, i.e., $E_f \simeq 1{\cdot}2$ eV for platinum.

Other applications for the instrument include a study of alloying effects, such as randomness and order in solid solutions, clustering at the early stages of precipitation, and the association of solute atoms with vacancies, dislocations, stacking faults and grain boundaries. Figure 113 shows a field-ion picture of vacancy-clusters in an equiatomic cobalt platinum alloy in the ordered state, and illustrates the type of defect structures found after ordering. The micrograph, which is a helium ion image taken at 77°K, was obtained from a tip radius of about 800 Å and it is deduced that the clusters are at the edges of a (100) plane.

NEUTRON DIFFRACTION

The advent of nuclear reactors has stimulated the application of neutron diffraction to those problems of physical metallurgy which cannot be solved satisfactorily by other diffraction techniques. A beam of neutrons from the reactor behaves as a wave motion, with a wavelength $\lambda = h/mv$, where m and v are the mass and velocity of the neutron. For a reactor which usually operates in the temperature range 0 to 100°C, the distribution of neutron wavelengths will be as shown schematically in Fig. 114. It contains only a continuous distribution of wavelengths, in contrast to the intense

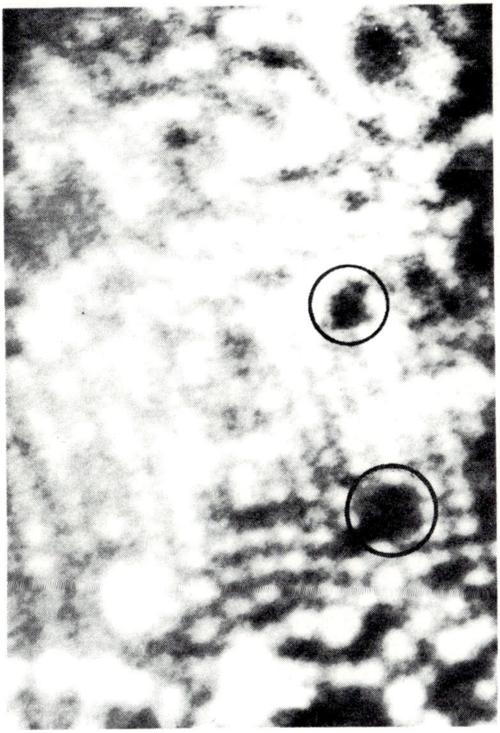

FIG. 113. Field-ion micrograph of equiatomic Co–Pt alloy, imaged with helium ions at 77°K, showing vacancy clusters. (After D. G. Brandon, *Brit. J. App. Phys.*, **14**, 474, 1963, fig. 96.)

monochromatic lines superimposed on a continuous background for X-rays. As a consequence, for most neutron diffraction work, it is necessary to separate off a narrow band of wavelengths by reflection from a lead crystal monochromator. The neutron beam is weaker than an X-ray beam by about 10^5 quanta per square cm per second and, to compensate for this, wide beams several inches in cross-section have to be used to achieve a sufficiently high counting rate in the $\mathbf{B^{10}F_3}$ proportional counter, some 18 in. long by 2 in. diameter. It therefore follows that neutron spectrometers, although similar in principle to X-ray spectrometers, have to be built on a massive scale.

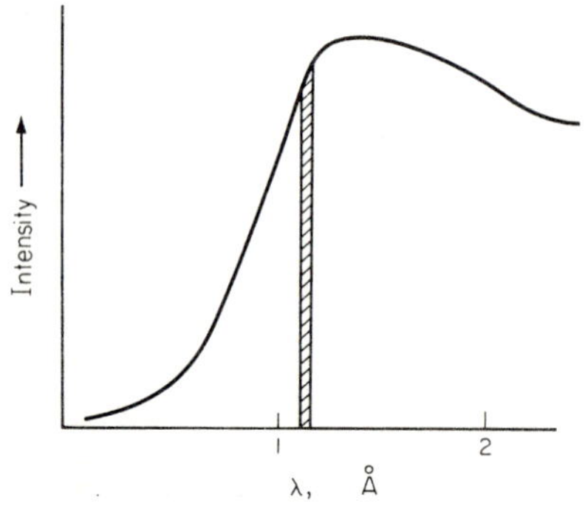

FIG. 114. Schematic diagram of neutron spectrum from a nuclear reactor.

It is possible to find out structural information with neutrons which cannot be obtained by X-rays, and this stems from the fact that, in general, it is the nucleus of the atom which scatters neutrons. Figure 115 shows how the scattering amplitude varies in quite a haphazard way for the elements of the periodic table. We see that the light elements, particularly hydrogen, scatter as strongly as heavier elements, and in contrast to X-ray studies when the weak intensity

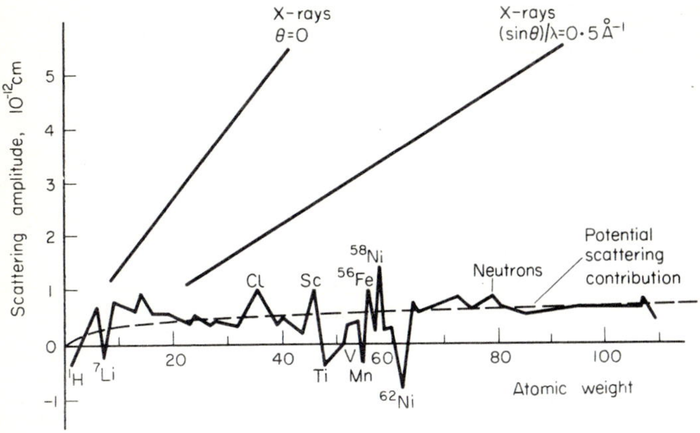

FIG. 115. The variation of scattering amplitude with atomic weight of the elements. (From *Research*, **7**, 257, 1954.)

contribution of the light atoms are swamped by those from the heavy atoms, in neutron work the scattering from light elements is easily detected. With neutrons it is possible to determine the positions of the oxygen ions in oxides of uranium, and the way in which the structure alters as the composition deviates from the stoichiometric value. We see, too, that neighbouring elements such as iron, nickel and cobalt, copper and zinc scatter quite differently and thus alloys made from neighbouring elements in the periodic table can be more easily studied with neutrons. One important application is in the study of ordered alloys and by using neutrons it has been possible to prove the existence of superlattice lines in the ordered alloys FeCo and Ni_3Mn. X-rays are useless in this type of study because the scattering powers of the different elements are practically identical. Conversely, neutrons are unable to show the superlattice lines in Cu_3Au although they are quite easy to detect using X-rays.

Such studies as those discussed above can be carried out by X-rays, but the problem is made easier with neutrons. Neutrons can be used, however, to study magnetic moments on an atomic scale and this application has no counterpart in X-ray diffraction studies. This application arises because neutrons are scattered by "magnetic atoms" (i.e., atoms having unpaired electron spins and hence a magnetic moment) since the neutron has a spin of its own. In a paramagnetic material there is no preferential alignment of the magnetic moments, so that these point randomly in all directions. The magnetic neutron scattering therefore shows up simply as an addition to the general background, decreasing in intensity with increasing angle. By measuring the increase in background intensity the magnitude of the magnetic moment may be deduced. In ferromagnetic and anti-ferromagnetic materials, more striking effects are found. In anti-ferromagnetic materials, the magnetic moments of neighbouring atoms are aligned in opposite directions, i.e., anti-parallel. A simple example is shown in Fig. 116, which is for the alloy

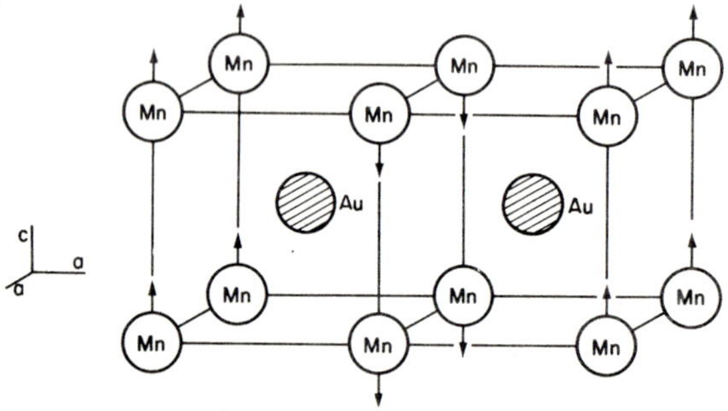

FIG. 116. The magnetic structure of AuMn at room temperature. The arrows represent the direction of the magnetic spins. (After G. Bacon, *Neutron Diffraction and Scattering*, A.E.R.E. M/R 2857, fig. 5, published by H.M.S.O.)

AuMn. From the point of view of the magnetic structure, the periodicity along the *a*-axis is twice that of the ordinary unit cell, and consequently lines, additional to those in the X-ray pattern, will appear in the neutron diffraction pattern which index as ($\frac{1}{2}$, 0, 0) ($\frac{1}{2}$, 1, 0), (3/2, 0, 0) etc. because the neutron scattering from the downward pointing magnetic moments is 180° out of phase with that from the upward moments; this effect is shown in Fig. 117(a). On heating the anti-ferromagnetic material, the regular arrangement of anti-parallel spins is destroyed and the additional "super-lattice" lines disappear (see Fig. 117 b). In ferromagnetic materials all the magnetic moments are aligned in the same direction within a domain, so that no superlattice lines appear in the diffraction pattern. There is, however, a strong magnetic contribution at the positions of the normal nuclear scattering, and this contribution will fall off as the temperature is raised to the Curie point.

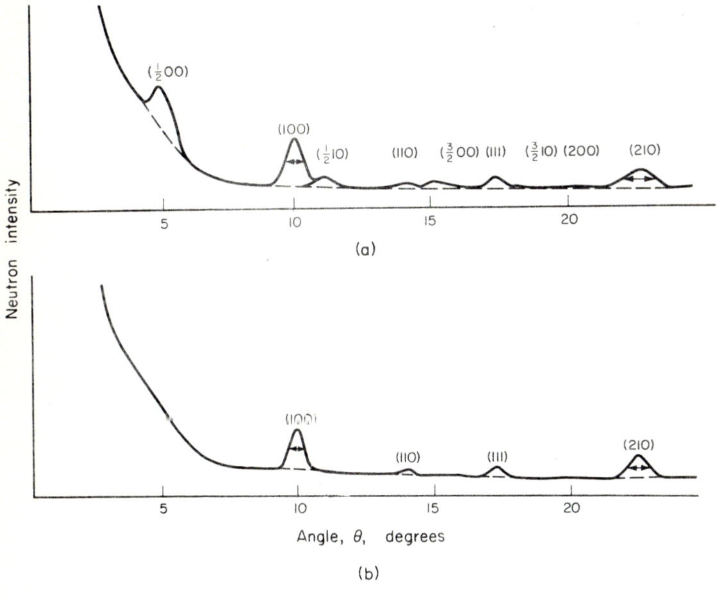

FIG. 117. Neutron diffraction pattern from AuMn (a) at room temperature showing the (½00) magnetic reflection and (b) at 260°C in the paramagnetic region. (After G. Bacon, *Neutron Diffraction and Scattering*, A.E.R.E. M/R 2857, fig. 6, published by H.M.S.O.)

Another property of neutrons which has not been exploited to any great extent is the low absorption property. This penetrative property of the neutron beam may well provide scope for future neutron crystallography, since it enables the whole body of a massive sample to be examined and not merely its surface. Problems concerned with preferred orientation and cavitation are possible applications.

Neutrons of wavelength longer than twice the longest lattice spacing may be used in special transmission studies. One successful application is in the field of small-angle scattering from lattice defects. The scattering from dislocations and the defects introduced by cold work, quenching or radiation damage is masked, to some extent, by the intensity scattered into the low angle region by two successive large-angle Bragg reflections; a ray reflected from one region of a grain suffers further reflection by another region in which the reflecting planes are somewhat misaligned by a simple rotation. By using neutrons of a wavelength greater than the Bragg cut-off wavelength (i.e., $\lambda = 2d_{max}$ where d_{max} is the maximum interplanar spacing), Bragg reflections cannot occur. It is therefore possible to measure the true small-angle scattering due to density changes associated with defects.

Long-wavelength X-rays are so heavily absorbed that the maximum specimen thickness which can be examined is only a few microns, whereas with long-wavelength neutrons, thick specimens ($\sim$1 cm) can be used. Such neutron studies have proved that there is no evidence for the existence of cavities in cold worked or fatigued metals.

SUGGESTIONS FOR FURTHER READING

E. W. MÜLLER, Field ionization and field ion microscopy, *Advances in Electronics and Electron Physics*, **13**, 83 (1960). Academic Press, New York.

Direct Observations of Imperfections in Crystals, Interscience Publication, 1963.

G. BACON, *Neutron Diffraction*.

INDEX